Advanced Spintronic and Electronic Nanomaterials

Advanced Spintronic and Electronic Nanomaterials

Gang Xiang
Hongtao Ren

MDPI

Basel • Beijing • Wuhan • Barcelona • Belgrade • Novi Sad • Cluj • Manchester

Gang Xiang
Physics
Sichuan University
Chengdu
China

Hongtao Ren
Materials Science and Engineering
Liaocheng University
Liaocheng
China

Editorial Office
MDPI AG
Grosspeteranlage 5
4052 Basel, Switzerland

This is a reprint of the Special Issue, published open access by the journal *Nanomaterials* (ISSN 2079-4991), freely accessible at: www.mdpi.com/journal/nanomaterials/special_issues/nano_spintronic.

For citation purposes, cite each article independently as indicated on the article page online and using the guide below:

Lastname, A.A.; Lastname, B.B. Article Title. *Journal Name* **Year**, *Volume Number*, Page Range.

ISBN 978-3-7258-1686-6 (Hbk)
ISBN 978-3-7258-1685-9 (PDF)
https://doi.org/10.3390/books978-3-7258-1685-9

Contents

About the Editors

Gang Xiang

Gang Xiang is a full professor at the College of Physics, Sichuan University, China. He earned his PhD in condensed matter physics from Pennsylvania State University, University Park, in 2006. His research interests include the design, fabrication, and characterization of semiconductors and magnetic materials, including diluted magnetic semiconductors and 2D materials such as transition metal chalcogenides, and their spintronic and electronic applications.

Hongtao Ren

Dr. Hongtao Ren is an instructor at the School of Materials Science and Engineering at Liaocheng University. He completed his PhD studies in Materials Science and Engineering at Xi'an Jiaotong University. His interests include straintronics, semiconductor spintronics, and 2D ferromagnetic materials.

nanomaterials

Editorial

Advanced Spintronic and Electronic Nanomaterials

Gang Xiang [1,*] and Hongtao Ren [2,*]

1 College of Physics, Sichuan University, Chengdu 610064, China
2 School of Materials Science and Engineering, Liaocheng University, Liaocheng 252059, China
* Correspondence: gxiang@scu.edu.cn (G.X.); renhongtao@lcu.edu.cn (H.R.)

Since single-layer graphene [1] with ultrahigh carrier mobility was obtained experimentally in 2004, two-dimensional (2D) layered electronic materials have become more widespread [2–9]. Two-dimensional non-layered materials [10–14], with their abundant terrestrial resources and low costs, support broader practical applications. Consequently, the repository of 2D materials has become more diverse, facilitating their application in spintronics [15–19], flexible electronics [20,21], information science [22,23], and related fields [24,25].

Over the past two decades, spintronics and electronics [26–30] have developed very rapidly. In 2017, low-temperature long-range ferromagnetic order was experimentally discovered both in $Cr_2Ge_2Te_6$ [26] and CrI_3 [27] monolayer systems. Two-dimensional ferromagnetism immediately became of tremendous interest to researchers all over the world. As such, studies on 2D materials have expanded and now correlate with investigations of both traditional materials and emerging materials including diluted magnetic semiconductors and wide band gap semiconductors.

This Special Issue brings together ten articles, specifically eight research articles and two review articles, dedicated to advanced spintronic and electronic nanomaterials. The content of the Special Issue includes the following: the modulation of vortex resonance in ferromagnetic permalloy dots [31], the capping layer effect on tunneling magnetoresistance in tunnel junctions [32], the size-dependent superconducting properties of indium nanowires [33], the co-doping effect of Mn and halogen elements on GeSe monolayers [34], the colossal magnetoresistance in layered diluted magnetic semiconductor $Rb(Zn,Li,Mn)_4As_3$ [35], charge density wave transitions in 2D 1T-TaS_2 crystals [36], characterizations of Mn_5Ge_3 contacts on Ge/SiGe heterostructures [37] and Ni-doped Cd_3As_2 films on GaAs (111) substrates [38], strain engineering of intrinsic ferromagnetism in 2D van der Waals materials [6], and spintronic applications of carbon-based nanomaterials [39]. Our Special Issue may promote and accelerate ongoing research efforts of advanced spintronic and electronic nanomaterials. It is of vital importance to 2D spintronic devices and will be of interest to general readers of Nanomaterials.

check for updates

Citation: Xiang, G.; Ren, H. Advanced Spintronic and Electronic Nanomaterials. *Nanomaterials* **2024**, *14*, 1139. https://doi.org/10.3390/nano14131139

Received: 17 June 2024
Revised: 27 June 2024
Accepted: 28 June 2024
Published: 2 July 2024

Author Contributions: H.R. and G.X. wrote this Editorial Letter. All authors have read and agreed to the published version of the manuscript.

Funding: H.R. acknowledges the Shandong Province Natural Science Foundation (Grant No. ZR202103040767). G.X. acknowledges the National Natural Science Foundation of China (NSFC) (Grant No. 52172272).

Acknowledgments: The Guest Editors thank the authors for submitting their work to the Special Issue and for its successful completion. A special thank you to all the reviewers participating in the peer-review process of the submitted manuscripts and for enhancing the papers' quality and impact. We are also grateful to thank all the staff in the Editorial Office who made the entire creation of the Special Issue a smooth and efficient process.

Conflicts of Interest: The authors declare no conflicts of interest.

References

1. Novoselov, K.S.; Geim, A.K.; Morozov, S.V.; Jiang, D.; Zhang, Y.; Dubonos, S.V.; Grigorieva, I.V.; Firsov, A.A. Electric Field Effect in Atomically Thin Carbon Films. *Science* **2004**, *306*, 666–669. [CrossRef] [PubMed]
2. Gibertini, M.; Koperski, M.; Morpurgo, A.F.; Novoselov, K.S. Magnetic 2D Materials and Heterostructures. *Nat. Nanotech.* **2019**, *14*, 408–419. [CrossRef] [PubMed]
3. Kurebayashi, H.; Garcia, J.H.; Khan, S.; Sinova, J.; Roche, S. Magnetism, Symmetry and Spin Transport in van der Waals Layered Systems. *Nat. Rev. Phys.* **2022**, *4*, 150–166. [CrossRef]
4. Chen, X.; Zhang, X.; Xiang, G. Recent Advances in Two-dimensional Intrinsic Ferromagnetic Materials Fe_3X (X = Ge and Ga) Te_2 and Their Heterostructures for Spintronics. *Nanoscale* **2023**, *16*, 527–554. [CrossRef] [PubMed]
5. Zhong, J.; Zhang, X.; He, W.; Gong, D.; Lan, M.; Dai, X.; Peng, Y.; Xiang, G. Large-scale Fabrication and Mo vacancy-induced Robust Room-temperature Ferromagnetism of $MoSe_2$ Thin Films. *Nanoscale* **2023**, *15*, 6844–6852. [CrossRef] [PubMed]
6. Ren, H.; Xiang, G. Strain Engineering of Intrinsic Ferromagnetism in 2D van der Waals Materials. *Nanomaterials* **2023**, *13*, 2378. [CrossRef] [PubMed]
7. Ren, H.; Xiang, G. Strain-modulated magnetism in MoS_2. *Nanomaterials* **2022**, *12*, 1929. [CrossRef] [PubMed]
8. Ren, H.; Xiang, G. Recent Progress in Research on Ferromagnetic Rhenium Disulfide. *Nanomaterials* **2022**, *12*, 3451. [CrossRef] [PubMed]
9. Ren, H.; Liu, Y.; Zhang, L.; Liu, K. Synthesis, Properties, and Applications of Large-scale Two-dimensional Materials by polymer-assisted deposition. *J. Semicond.* **2019**, *40*, 061003. [CrossRef]
10. Ren, H.; Xiang, G. Recent Advances in Synthesis of Two-Dimensional Non-van Der Waals Ferromagnetic Materials. *Mater. Today Electron.* **2023**, *6*, 100074. [CrossRef]
11. Fan, X.; Chen, Z.; Xu, D.; Zou, L.; Ouyang, F.; Deng, S.; Wang, X.; Zhao, J.; Zhou, Y. Phase-controlled Synthesis of Large-area Trigonal 2D Cr_2S_3 Thin Films via Ultralow Gas-Flow Governed Dynamic Transport. *Adv. Funct. Mater.* **2024**, *34*, 2404750. [CrossRef]
12. Dai, X.; Zhang, X.; Gong, D.; Xiang, G. Performance Enhancement and In Situ Observation of Resistive Switching and Magnetic Modulation by a Tunable Two-Level System of Mn Dopants in a-Gallium Oxide-based Memristor. *Adv. Funct. Mater.* **2023**, *33*, 2304749. [CrossRef]
13. Boi, F.S.; Guo, J.; Xiang, G.; Lan, M.; Wang, S.; Wen, J.; Zhang, S.; He, Y. Cm-size Free-standing Self-organized Buckypaper of Bucky-onions Filled with Ferromagnetic Fe_3C. *RSC Adv.* **2017**, *7*, 845–850. [CrossRef]
14. Ren, H.; Xiang, G.; Gu, G.; Zhang, X.; Wang, W.; Zhang, P.; Wang, B.; Cao, X. Zinc Vacancy-Induced Room-Temperature Ferromagnetism in Undoped ZnO Thin Films. *J. Nanomater.* **2012**, *1*, 295358. [CrossRef]
15. Žutić, I.; Fabian, J.; Das Sarma, S. Spintronics: Fundamentals and Applications. *Rev. Mod. Phys.* **2004**, *76*, 323–410. [CrossRef]
16. Burch, K.S.; Mandrus, D.; Park, J.G. Magnetism in Two-Dimensional Van Der Waals Materials. *Nature* **2018**, *563*, 47–52. [CrossRef] [PubMed]
17. Ren, H.; Xiang, G.; Lu, J.; Zhang, X.; Zhang, L. Biaxial Strain-mediated Room Temperature Ferromagnetism of ReS_2 Web Buckles. *Adv. Electron. Mater.* **2019**, *5*, 1900814. [CrossRef]
18. Ren, H.; Zhang, L.; Xiang, G. Web Buckle-mediated Room-temperature Ferromagnetism in Strained MoS_2 Thin Films. *Appl. Phys. Lett.* **2020**, *116*, 012401. [CrossRef]
19. Ren, H.; Lan, M. Progress and Prospects in Metallic Fe_xGeTe_2 ($3 \leq x \leq 7$) Ferromagnets. *Molecules* **2023**, *28*, 7244. [CrossRef]
20. Liu, A.; Zhang, X.; Liu, Z.; Li, Y.; Peng, X.; Li, X.; Qin, Y.; Hu, C.; Qiu, Y.; Jiang, H.; et al. The Roadmap of 2D Materials and Devices toward Chips. *Nano-Micro Lett.* **2024**, *16*, 119. [CrossRef]
21. Cao, W.; Bu, H.; Vinet, M.; Cao, M.; Takagi, S.; Hwang, S.; Ghani, T.; Banerjee, K. The Future Transistors. *Nature* **2023**, *620*, 501–515. [CrossRef] [PubMed]
22. Qiu, H.; Yu, Z.; Zhao, T.; Zhang, Q.; Xu, M.; Li, P.; Li, T.; Bao, W.; ChaiI, Y.; Chen, S.; et al. Two-dimensional Materials for Future Information Technology: Status and Prospects. *Sci. China Inf. Sci.* **2024**, *67*, 160400. [CrossRef]
23. Zeng, S.; Liu, C.; Zhou, P. Transistor Engineering Based on 2D Materials in the Post-Silicon Era. *Nat. Rev. Electr. Eng.* **2024**, *1*, 335–348. [CrossRef]
24. Yi, H.; Ma, Y.; Ye, Q.; Lu, J.; Wang, W.; Zheng, Z.; Ma, C.; Yao, J.; Yang, G. Promoting 2D Material Photodetectors by Optical Antennas Beyond Noble Metals. *Adv. Sensor Res.* **2023**, *2*, 2200079. [CrossRef]
25. Samizadeh Nikoo, M.; Matioli, E. Electronic Metadevices for Terahertz Applications. *Nature* **2023**, *614*, 451–455. [CrossRef] [PubMed]
26. Gong, C.; Li, L.; Li, Z.L.; Ji, H.W.; Stern, A.; Xia, Y.; Cao, T.; Bao, W.; Wang, C.Z.; Wang, Y.A.; et al. Discovery of Intrinsic Ferromagnetism in Two-dimensional Van der Waals Crystals. *Nature* **2017**, *546*, 265–269. [CrossRef] [PubMed]
27. Huang, B.; Clark, G.; Navarro-Moratalla, E.; Klein, D.R.; Cheng, R.; Seyler, K.L.; Zhong, D.; Schmidgall, E.; McGuire, M.A.; Cobden, D.H.; et al. Layer-dependent Ferromagnetism in a Van der Waals Crystal Down to the Monolayer Limit. *Nature* **2017**, *546*, 270–273. [CrossRef]
28. Boi, F.S.; Guo, J.; Xiang, G.; Lan, M.; Wang, S.; Wen, J.; Zhang, S.; He, Y. Controlling the Quantity of α-Fe Inside Multiwall Carbon Nanotubes Filled with Fe-based Crystals: The Key Role of Vapor Flow-rate. *Appl. Phys. Lett.* **2014**, *105*, 243108. [CrossRef]
29. Wang, H.; Sun, S.; Lu, J.; Xu, J.; Lv, X.; Peng, Y.; Zhang, X.; Wang, Y.; Xiang, G. High Curie Temperature Ferromagnetism and High Hole Mobility in Tensile Strained Mn-Doped SiGe Thin Films. *Adv. Funct. Mater.* **2020**, *30*, 2002513. [CrossRef]

30. Feng, Y.; Zhang, X.; Zhao, G.; G Xiang, G. A Skyrmion Diode Based on Skyrmion Hall Effect. *IEEE Trans. Electron. Devices* **2022**, *69*, 1293–1297. [CrossRef]
31. Hu, S.; Cui, X.; Wang, K.; Yakata, S.; Kimura, T. Significant Modulation of Vortex Resonance Spectra in a Square-Shape Ferromagnetic Dot. *Nanomaterials* **2022**, *12*, 2295. [CrossRef]
32. Kim, G.; Lee, S.; Lee, S.; Song, B.; Lee, B.-K.; Lee, D.; Lee, J.S.; Lee, M.H.; Kim, Y.K.; Park, B.-G. The Influence of Capping Layers on Tunneling Magnetoresistance and Microstructure in CoFeB/MgO/CoFeB Magnetic Tunnel Junctions upon Annealing. *Nanomaterials* **2023**, *13*, 2591. [CrossRef]
33. Noyan, A.A.; Ovchenkov, Y.A.; Ryazanov, V.V.; Golovchanskiy, I.A.; Stolyarov, V.S.; Levin, E.E.; Napolskii, K.S. Size-Dependent Superconducting Properties of In Nanowire Arrays. *Nanomaterials* **2022**, *12*, 4095. [CrossRef]
34. He, W.; Zhang, X.; Gong, D.; Nie, Y.; Xiang, G. Mn-X (X = F, Cl, Br, I) Co-Doped GeSe Monolayers: Stabilities and Electronic, Spintronic and Optical Properties. *Nanomaterials* **2023**, *13*, 1862. [CrossRef]
35. Peng, Y.; Shi, L.; Zhao, G.; Zhang, J.; Zhao, J.; Wang, X.; Deng, Z.; Jin, C. Colossal Magnetoresistance in Layered Diluted Magnetic Semiconductor $Rb(Zn,Li,Mn)_4As_3$ Single Crystals. *Nanomaterials* **2024**, *14*, 263. [CrossRef]
36. Pan, X.; Yang, T.; Bai, H.; Peng, J.; Li, L.; Jing, F.; Qiu, H.; Liu, H.; Hu, Z. Controllable Synthesis and Charge Density Wave Phase Transitions of Two-Dimensional 1T-TaS_2 Crystals. *Nanomaterials* **2023**, *13*, 1806. [CrossRef]
37. Hutchins-Delgado, T.A.; Addamane, S.J.; Lu, P.; Lu, T.-M. Characterization of Mn_5Ge_3 Contacts on a Shallow Ge/SiGe Heterostructure. *Nanomaterials* **2024**, *14*, 539. [CrossRef]
38. Liang, G.; Zhai, G.; Ma, J.; Wang, H.; Zhao, J.; Wu, X.; Zhang, X. Circular Photogalvanic Current in Ni-Doped Cd_3As_2 Films Epitaxied on GaAs(111)B Substrate. *Nanomaterials* **2023**, *13*, 1979. [CrossRef]
39. Pawar, S.; Duadi, H.; Fixler, D. Recent Advances in the Spintronic Application of Carbon-Based Nanomaterials. *Nanomaterials* **2023**, *13*, 598. [CrossRef]

nanomaterials

Review

Strain Engineering of Intrinsic Ferromagnetism in 2D van der Waals Materials

Hongtao Ren [1,*] and Gang Xiang [2,*]

1 School of Materials Science and Engineering, Liaocheng University, Hunan Road No. 1, Liaocheng 252000, China
2 College of Physics, Sichuan University, Wangjiang Road No. 29, Chengdu 610064, China
* Correspondence: renhongtao@lcu.edu.cn (H.R.); gxiang@scu.edu.cn (G.X.)

Abstract: Since the discovery of the low-temperature, long-range ferromagnetic order in monolayers $Cr_2Ge_2Te_6$ and CrI_3, many efforts have been made to achieve a room temperature (RT) ferromagnet. The outstanding deformation ability of two-dimensional (2D) materials provides an exciting way to mediate their intrinsic ferromagnetism (FM) with strain engineering. Here, we summarize the recent progress of strain engineering of intrinsic FM in 2D van der Waals materials. First, we introduce how to explain the strain-mediated intrinsic FM on Cr-based and Fe-based 2D van der Waals materials through ab initio Density functional theory (DFT), and how to calculate magnetic anisotropy energy (MAE) and Curie temperature (T_C) from the interlayer exchange coupling J. Subsequently, we focus on numerous attempts to apply strain to 2D materials in experiments, including wrinkle-induced strain, flexible substrate bending or stretching, lattice mismatch, electrostatic force and field-cooling. Last, we emphasize that this field is still in early stages, and there are many challenges that need to be overcome. More importantly, strengthening the guideline of strain-mediated FM in 2D van der Waals materials will promote the development of spintronics and straintronics.

Keywords: strain engineering; ferromagnetism; transition metal trihalides; transition metal chalcogenides; transition metal phosphorous chalcogenides; wrinkle; flexible substrates; lattice mismatch; spintronics; straintronics

Citation: Ren, H.; Xiang, G. Strain Engineering of Intrinsic Ferromagnetism in 2D van der Waals Materials. *Nanomaterials* **2023**, *13*, 2378. https://doi.org/10.3390/nano13162378

Academic Editor: Stefano Agnoli

Received: 9 July 2023
Revised: 9 August 2023
Accepted: 17 August 2023
Published: 19 August 2023

1. Introduction

According to the *Mermin–Wanger–Hohenberg* theory [1,2], thermal fluctuations can destroy the long-range magnetic order of 2D systems at finite temperatures. However, the anisotropy of the system suppresses thermal disturbances by opening the gap in the spin-wave spectrum [3–6]. Furthermore, spin orbit coupling (SOC) can stabilize the long-range magnetic order in 2D systems by contributing to magnetic anisotropy. After the discovery of the low-temperature, long-range ferromagnetic order in monolayers $Cr_2Ge_2Te_6$ and CrI_3 [7,8], many efforts have been made to achieve a room temperature (RT) ferromagnet. Indeed, strain engineering [9–25] is a very important strategy for mediating material properties, including optoelectronic [9,10,13–15,21], electrocatalytic [11,16,22–24], and magnetic properties [15,19,21,24–28]. Since Novoselov et al. [29] obtained a stable monolayer graphene in the laboratory in 2004, further research gradually revealed that 2D materials, such as MoS_2, could withstand up to 20% strain [30–33]. However, it was very difficult to directly apply strain to 2D materials in experiments, which made strain-controlled performance largely remain in theoretical study. This was because by changing lattice parameters, strain could be easily applied to the lattice of 2D materials. Specifically, the study on strain-mediated magnetism in 2D materials, especially in 2D van der Waals materials with intrinsic long-range FM order, was focused on theoretical calculation. More interestingly, the $Bi_2Te_3 | Fe_3GeTe_2$ heterostructure related to strain [34,35] was designed to increase the Curie temperature (T_C) due to the proximity effect [36–40].

Very recently, some significant progress was also made in the field of experimental research [27,28,41–48]. In Figure 1, we summarize the recent progress of strain engineering of intrinsic ferromagnetism (FM) in 2D van der Waals materials. First, we introduce how to explain the strain-mediated intrinsic FM on Cr-based and Fe-based 2D van der Waals materials with long-range FM order through ab initio Density functional theory (DFT), and how to calculate magnetic anisotropy energy (MAE) and Curie temperature from the interlayer exchange coupling J. Subsequently, we focus on numerous attempts to apply strain to 2D materials in experiments, including wrinkle-induced strain, flexible substrate bending or stretching, lattice mismatch, electrostatic force and field-cooling. Last, we emphasize that this field is still in early stages, and there are many challenges that must be overcome. More importantly, strengthening the guideline of strain-mediated FM in 2D van der Waals materials will promote the development of spintronics [6,49–57] and straintronics [12,19]. As shown in Figure 1, we summarize three different kinds of 2D materials with intrinsic long-range FM order, including transition metal trihalides ($CrCl_3$, $CrBr_3$ and CrI_3) [5,58–72], transition metals chalcogenides ($Cr_2Ge_2Te_6$, Fe_nGeTe_2 and $CrTe_2$) [42–46,73–87], and transition metal phosphorous chalcogenides ($AgVP_2Se_6$, $CrPS_4$) [88–91].

Figure 1. The three different kinds of 2D van der Waals materials with intrinsic long-range FM order.

2. Theoretical Calculations

In order to understand the essence of 2D ferromagnetism, ab initio Density functional theory, including linear density approach (LDA) [59], local spin density approximation (LSDA) [92], the generalized gradient approach (GGA) [61,93,94], and DFT + U [95,96], was often used to calculate the electronic structure of the system as a starting point. Moreover, the interlayer exchange coupling J was closely related to magnetic anisotropy, and it would also be used to calculate T_C [86,97–100]. The mean field theory [97,101] would roughly estimate T_C but, often, T_C overestimated it. Although the random phase approximation (RPA) could more accurately estimate T_C of three-dimensional (3D) materials, it may fail in 2D systems with large anisotropy. Notably, classic Monte Carlo (MC) [6,97] simulations can also describe the critical temperature.

2.1. Cr-Based 2D van der Waals Materials

2.1.1. $CrCl_3$

Unlike bulk materials, 2D materials can sustain larger strains [33,102]. Similarly, single-layer transition metal trihalides (MX_3, I, Cl and Br) can also withstand a strain of about 10% [58]. As a typical example, Yan et al. [59] studied the biaxial strain dependence magnetic anisotropy energy (MAE) of the 2D monolayer $CrCl_3$ (Figure 2a–d). When the compressive strain reached 2.5%, a phase transition from antiferromagnetism (AFM) to FM occurred (Figure 2e). In addition, when tensile strain was 2.4%, the maximum Curie temperature (T_C) reached 39 K. The MAE in the unstrained monolayer was positive, indicating the spins of Cr atoms were off-plane (Figure 2f).

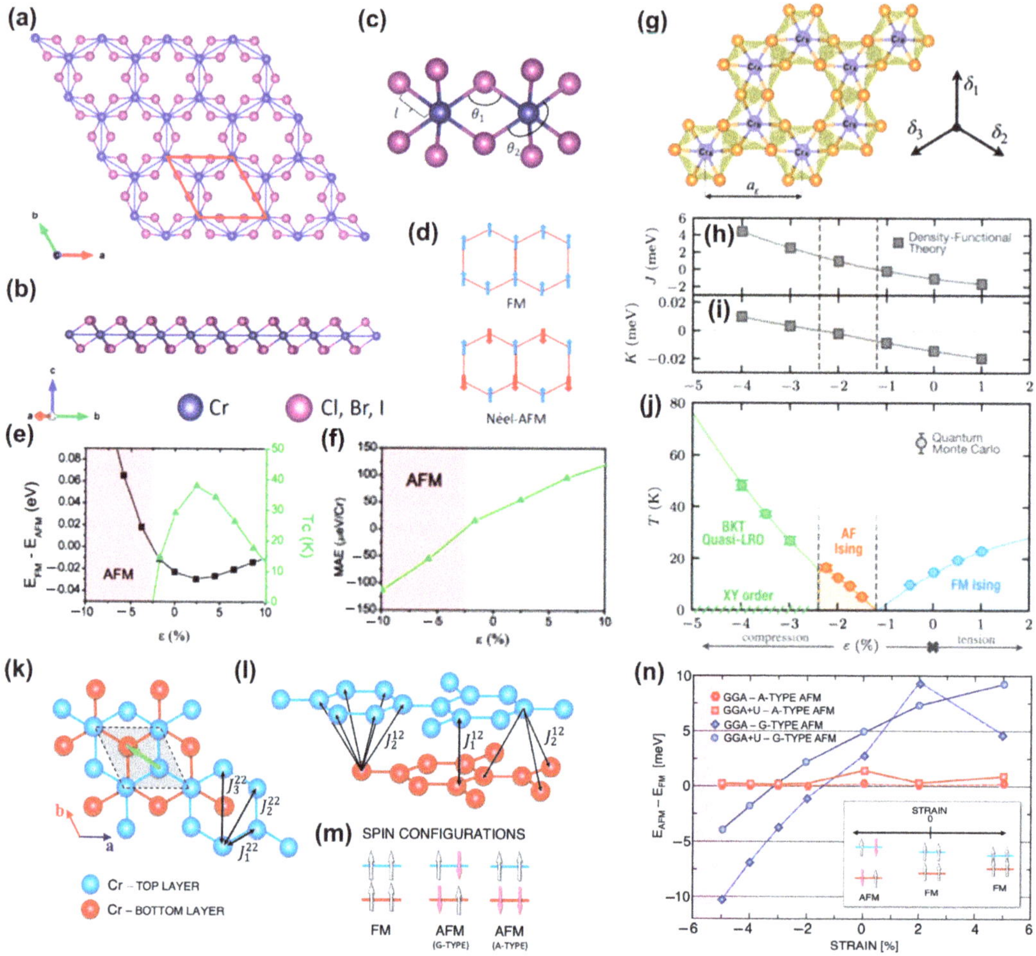

Figure 2. (**a**) Top and (**b**) side view of atomic structure of monolayer CrX_3 (X = Cl, Br, I). (**c**) Bonding between chromium and iodine atoms. The unit cell of CrX_3, which includes two Cr and six X atoms, has been indicated in (**a**). The bond length between Cr and an X atom, the bond angle θ_1 between Cr and two X atoms in the same plane, and the axial angle θ_2 are also shown in (**c**). (**d**) The two magnetic orders, namely AFM and FM. Energy difference between the FM and AFM phases for (**e**) $CrCl_3$. The AFM phase region is highlighted in red. The calculated T_C is also shown for each case. Change in MAE with respect to strain in (**f**) $CrCl_3$. (Reproduced with permission from [59]. Copyright 2016, American Physical Society). (**g**) Crystal structure of monolayer $CrCl_3$. Dashed lines denote the unit cell with basic vectors $\delta_1 = a_\varepsilon(0,1), \delta_2 = a_\varepsilon\left(\sqrt{3}/2,\ -1/2\right), \delta_3 = a_\varepsilon\left(-\sqrt{3}/2, -1/2\right)$, with strain-dependent lattice constanta a_ε. (**h**,**i**) Magnetic nearest-neighbor superexchange J and anisotropy K of Hamiltonian (1), respectively, computed via Density Functional Theory (DFT) as a function of monolayer strain ε. (**j**) Finite−temperature phase diagram of the monolayer $CrCl_3$ versus strain ε. Strain drives the monolayer into three different finite−temperature magnetic phases: BKT quasi-LRO phase, AFM Ising, and FM Ising. (Reproduced with permission from [63]. Copyright 2021, American Physical Society). (**k**,**l**) Arrangement of Cr atoms in the lattice of the $CrCl_3$ bilayer in the low temperature phase. The green arrow indicates the direction of lateral shift between the top and bottom layers. (**m**) Schematic plot of three different spin configurations of bilayer $CrCl_3$. (**n**) The energy difference between the FM and AFM phases. (Reproduced with permission from [64]. Copyright 2023, Springer Nature).

After applying the strain to the lattice, Cl atoms adjusted their position to minimize the lattice distortion energy at this strain (Figure 2g). After generating a specific structure, the energy difference between the ferromagnetic and antiferromagnetic states was calculated. Mapping this energy difference to the Hamiltonian (1), Dupont et al. [63] obtained the nearest-neighbor exchange coupling J and magnetic anisotropy K, as shown in Figure 2h,i.

$$\hat{H} = J_{\varepsilon} \sum_{\langle \vec{r}, \vec{r}' \rangle} \vec{S}_{\vec{r}} \cdot \vec{S}_{\vec{r}'} + K_{\varepsilon} \sum_{\vec{r}} \left(\vec{S}^{z}_{\vec{r}} \right)^2 \quad (1)$$

Note that: the spin value of S was 3/2 in the above equation.

As the strain evolved from compression to tension, the system sequentially exhibited the BKT (Berezinskii–Kosterlitz–Thouless) quasi-long-range order (LRO) phase, AFM Ising and FM Ising by QMC (Quantum Monte Carlo) simulations. Although theoretical and experimental studies have been conducted on monolayer and bulk $CrCl_3$ materials, research on their multilayer structures, including bilayers and trilayers systems, was very limited. Ebrahimian et al. [64] found that biaxial strains could also achieve a phase transition from AFM to FM (Figure 2k–n). In addition, the magnetic anisotropy could be mediated by the strain.

2.1.2. $CrBr_3$

More interestingly, the Curie temperature of monolayer $CrBr_3$ could be increased to 314 K by doping [103], which was between $CrCl_3$ (323 K) and CrI_3 (293 K). Although both hole doping and electron doping could enhance ferromagnetic coupling, the effect of hole doping was better at the same doping concentration. After applying biaxial strain to a unit cell, its magnetic moment remained unchanged, which indicated that the biaxial strain could not effectively enhance the ferromagnetic coupling of monolayer CrX_3. In addition, Webster et al. [59] found that applying a tensile strain of 2.1% could increase the T_C to 44 K, which was about 5 K higher than when no strain was applied. However, at a compressive strain of −4.1%, a FM to AFM phase transition, similar to $CrCl_3$ [59,63,64] and CrI_3 [59,70–72], also occurred.

2.1.3. CrI_3

Unlike monolayer $CrCl_3$, the electronic bandgap of monolayer CrI_3 remained almost unchanged after applying biaxial tensile strain; after applying biaxial compressive strain, the electronic bandgap decreased significantly and MAE increased significantly. When compressive strain reached 5%, MAE increased by 47% [59]. Continuing to increase the strain (−5.7%) resulted in a phase transition from FM to AFM. Similar to $CrCl_3$, Wu et al. [69] also found that the CrI_3 monolayer underwent a complex phase transition from magnetic metals, half-metal, half-semiconductor to magnetic semiconductor as the strain evolved from compression (−15%) to tension (10%).

As a typical example, Vishkayi et al. [71] investigated the effects of biaxial and uniaxial strain on the magnetism of monolayer CrI_3 (Figure 3a–e). A similar phase transition from FM to AFM [59] was also observed when a compressive strain greater than 7% was applied (Figure 3d). As the strain increased, the electronic bandgap showed an opposite trend when applying compressive or tensile strain (Figure 3d). Interestingly, uniaxial strain enhanced the nearest neighbor, Dzialoshinskii–Moriya (DM) interaction, by breaking the inversion symmetry, and its effect was stronger than biaxial strain.

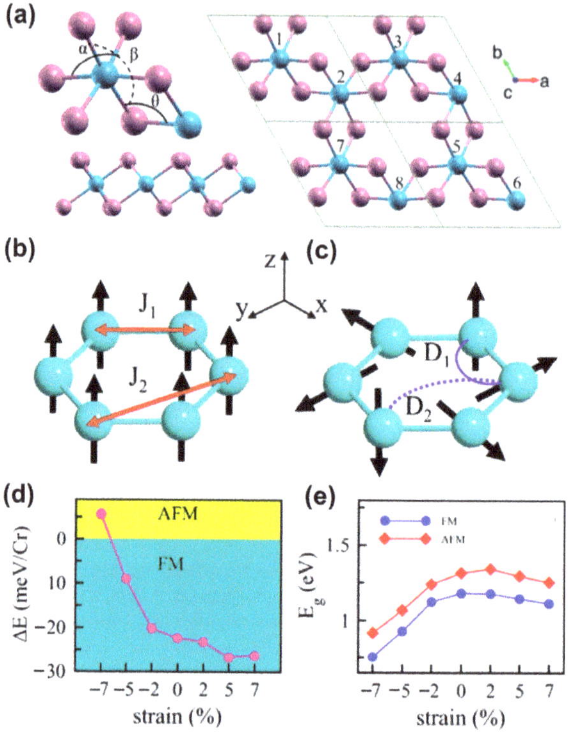

Figure 3. (**a**) Top and side views of monolayer CrI_3. The bonding angles between the atoms in the monolayer are denoted by θ, α, and β. A 2 × 2 × 1 supercell of the monolayer. The blue (pink) spheres represent Cr(I) atoms. (**b**) Schematic picture of the symmetric exchange couplings between Cr atoms, where J_1 denotes the coupling between nearest-neighbor atoms and J_2 denotes the coupling between next-nearest-neighbor atoms. (**c**) The same as (**b**) for the DM vectors. (**d**) The total energy difference between the FM and AFM configurations, ΔE, and (**e**) the variation in the band gap as a function of strain for monolayer CrI_3 in the FM and AFM configurations. (Reproduced with permission from [71]. Copyright 2020, American Physical Society).

The phase transition from FM to AFM under compressive strain (−3% or −5%) had also been discovered in the CrI_3 bilayer [70], similar to other previously reported systems [59,71]. In addition, Safi et al. [72] found that the phase transition occurred at −6% compressive strain. More importantly, they also discovered a second phase transition point from FM to AFM, which occurred near −2.5% compressive strain.

2.1.4. $CrTe_2$

Guo et al. [104]. found that applying uniaxial or biaxial tensile strain to monolayer $CrTe_2$ in the *T* phase did not cause a phase transition from direct to indirect bandgap. Under biaxial strain, the $CrTe_2$ monolayer [105] underwent phase transformation at −1% compressive strain in Figure 4a–k.

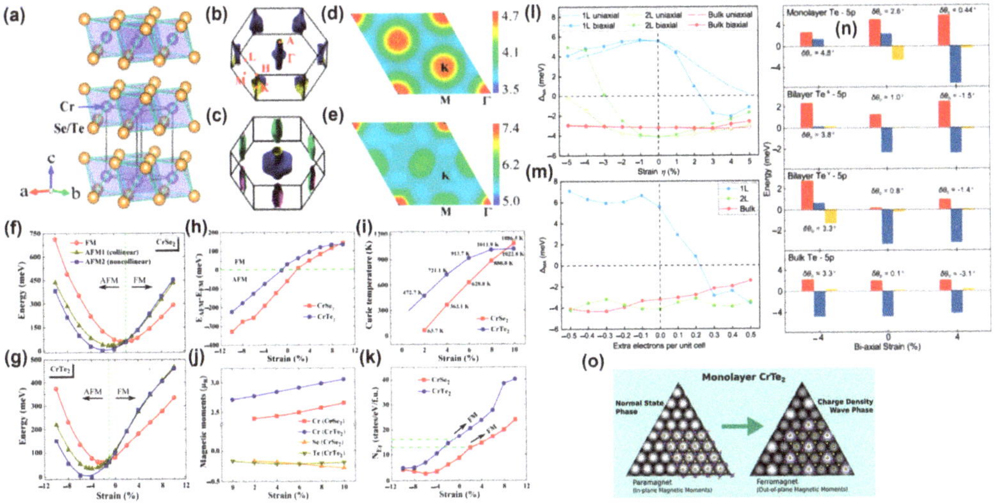

Figure 4. (**a**) Crystal structure of $CrSe_2$/$CrTe_2$ bulk; Fermi surface without spin polarization for (**b**) $CrSe_2$ and (**c**) $CrTe_2$ bulks; real part of electron susceptibility χ' with $q_z = 0$ for (**d**) $CrSe_2$ and (**e**) $CrTe_2$ bulks. Relative total energies of three different magnetic configurations [FM, AFM1 (collinear), and AFM2 (non-collinear)] as a function of biaxial strain for (**f**) $CrSe_2$ and (**g**) $CrTe_2$ monolayers. Strain dependence of (**h**) the energy difference ΔE ($=E_{AFM} - E_{FM}$) between AFM and FM states in one unit cell, (**i**) the T_C in FM states, (**j**) magnetic moments on Cr and Se/Te atoms in FM states, and (**k**) the number of density of states (DOS) at the Fermi energy N_{EF} in the nonmagnetic (NM) states for $CrSe_2$/$CrTe_2$ monolayers. (Reproduced with permission from [105]. Copyright 2015, American Physical Society). (**l**) strain and (**m**) band filling of 1 L, 2 L, and bulk 1T-$CrTe_2$. Difference in SOC matrix elements $\Delta_{<pi|pj>}$ (per atom) of the Te-5p orbitals. (**n**) $\Delta_{<pi|pj>}$ of 1 L, 2 L, and bulk 1T-$CrTe_2$ versus strain. At each strain, the values for $\delta\theta_D$ are also shown. Positive and negative values of the strain correspond to tensile and compressive strain, respectively. For the bilayer, values for Te atoms at the van der Waals gap (Te^V) and Te atoms at the free surface Te^S are shown. (Reproduced with permission from [106]. Copyright 2022, American Physical Society). (**o**) Computed STM images for monolayer $CrTe_2$ in the normal state and the charge density wave (CDW) phase. (Reproduced with permission from [107]. Copyright 2020, American Chemical Society).

Interestingly, the $CrTe_2$ monolayer with strain-free was a FM state. After the tensile strain was applied, its T_C could rise to 1022.8 K [106], and the magnetic moment of Cr atom increased linearly, which may be caused by the increase of the density of states at Fermi energy N_{EF}. Magnetic anisotropy exhibited a different sensitivity to uniaxial and biaxial strain, as shown in Figure 4l. A monolayer was more sensitive to tensile strain, while a bilayer was more sensitive to compressive strain, and bulk was insensitive to the applied strain. Furthermore, band filling [108] of the monolayer also underwent a transition from out-of-plane to in-plane, while the bilayer and bulk did not undergo this transition (Figure 4m). More interestingly, the strain had the greatest impact on the dihedral angle θ_D, followed by the effect of Cr-Te-Cr bond length (Figure 4n). Notably, the charge density wave (CDW) phase [107] promoted greater stability of the long-range FM order.

2.1.5. $Cr_2Ge_2Te_6$

As early as 2014, Li et al. [73] predicted that $Cr_2X_2Te_6$ (X = Si, Ge), a layered crystal with intrinsic FM, could be obtained experimentally by exfoliation. Furthermore, the T_C of $Cr_2Ge_2Te_6$ ($Cr_2Si_2Te_6$) was increased by 51.7(56) K under an isotropic tensile strain. Similarly, Dong et al. [75] also found that $Cr_2Ge_2Se_6$ and $Cr_2Ge_2Te_6$ were a stable FM semiconductor, as shown in Figure 5a–e.

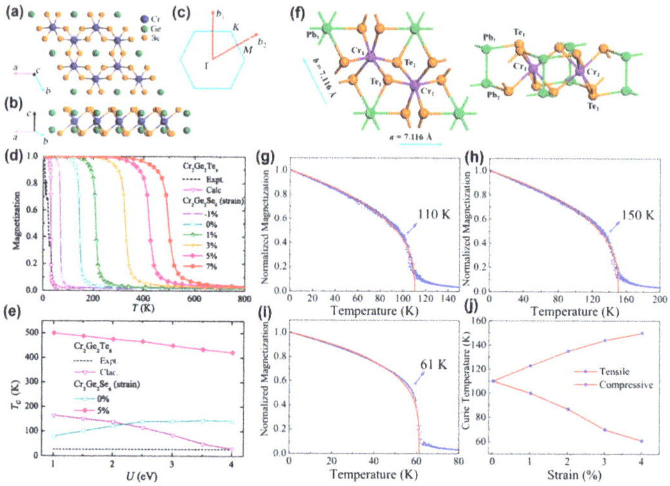

Figure 5. (**a**) top view in the *a-b* plane and (**b**) side view in the *a-c* plane of crystal structure of $Cr_2Ge_2Se_6$. The 2D Brillouin zone is shown in (**c**). (**d**) *M-T*. The experimental results for are taken from Ref. [7]. The calculated results are obtained by DFT calculations and Monte Carlo simulations. (**e**) T_C-*U*. (Reproduced with permission from [75]. Copyright 2019, American Physical Society). (**f**) top and side view of crystal structure of $CrPbTe_3$. *M-T* curves for (**g**) pristine $CrPbTe_3$, (**h**) 4% tensile strain structure, (**i**) 4% compressive strain structure and (**j**) variation of T_C with strain. (Reproduced with permission from [78]. Copyright 2020, IOP Publishing).

Through Monte Carlo simulation, it was found that the T_C of $Cr_2Ge_2Se_6$ (144 K) was five times that of $Cr_2Ge_2Te_6$ (30 K) (Figure 5d). Interestingly, when compressive strain was applied, the T_C decreased, and even at 2% strain, the phase became AFM. After applying tensile strain, the T_C would be increased, even up to 500 K. Notably, the T_C of $Cr_2Ge_2Se_6$ with 5% strain was always higher than RT (Figure 5e). To our knowledge, 2D $Cr_2Ge_2Se_6$ has not yet been prepared experimentally.

As a member of Cr-based materials, 2D $CrPbTe_3$ (CPT) has a stable monolayer structure and a higher T_C, as shown in Figure 5f. Similar to $Cr_2Ge_2Se_6$, its T_C (110 K) gradually increased from compressive strain (61 K) to tensile strain (150 K) (Figure 5g–j). In addition, strain can also induce spin reorientation from the in-plane to the out-of-plane.

2.1.6. $CrPS_4$

$AgVP_2Se_6$, as a typical quaternary FM semiconductor, was synthesized by heating the elements in evacuated silica tubes as early as 1988 [88]. Ouvrard et al. found that its polycrystalline powder showed low-temperature FM, high-temperature PM and T_C of about 29 K. Recently, single crystal $AgVP_2Se_6$ samples [90] exhibited better stability than MX_3 materials. As another typical example, single crystal $CrPS_4$ [91] was obtained by the chemical vapor transport method and mechanical exfoliation. Interestingly, odd layers were ferromagnetic at lower temperatures, while even layers were antiferromagnetic, as shown in Figure 6. To the best our knowledge, there were currently no reports on regulating the magnetic properties of $AgVP_2Se_6$ and $CrPS_4$ through strain engineering.

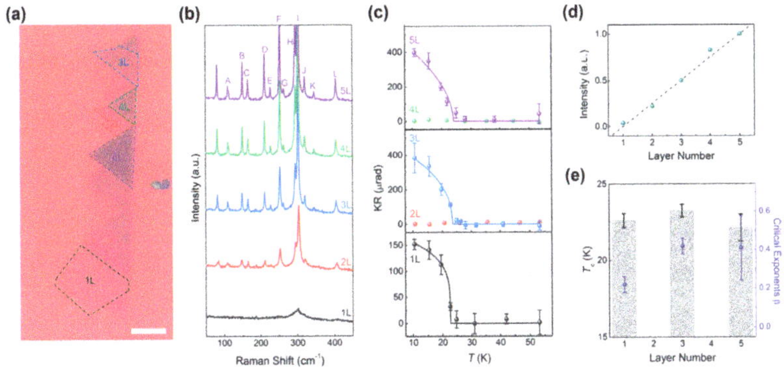

Figure 6. Layer dependent magnetic properties of thin $CrPS_4$. (**a**) Optical image. Scale bar is 5 μm. (**b**) Raman spectra. (**c**) Temperature dependence of MOKE signal under $\mu_0H = 0.25$ T measured on 1 L to 5 L flakes shown in (**a**). (**d**) Raman peak intensity of the F peak for each flake normalized by that from the 5 L as a function of the layer number. (**e**) Critical temperature, T_C (black dashed bar), and critical exponent, β (blue symbol), extracted from the fitting lines shown in (**c**). (Reproduced with permission from [68]. Copyright 2021, American Chemical Society).

2.2. Fe-Based 2D van der Waals Materials

The bulk Fe_3GeTe_2 (FGT) with the hexagonal platelets was first synthesized by a direct solid-state reaction (SSR) [81]. Zhuang et al. [82] found that applying biaxial strain on single-layer metallic Fe_3GeTe_2 could enhance the MAE and total magnetic moment (Figure 7a–d). When 2% tensile strain was applied, its MAE increased by 50%; After the applied strain changed from compression to tension, the total magnetic moment also increased with the increase of strain. Through DFT calculations, Hu et al. [85] discovered that monolayer FGT was a metallic type ferromagnet, with its magnetic moment mainly localized on Fe atoms and its metallicity mainly derived from the Fe *d* orbitals in Figure 7e–g.

After applying the biaxial strain shown in Figure 7h, there was a significant change in the spin-density distributions of the FGT monolayer. The magnetism of this system mainly came from Fe atoms (Figure 7i), especially Fe3 atoms at the inequivalent site (Figure 7f), with little contribution from Ge and Te atoms (Figure 7j). Due to the Poisson effect, when stretched in the in-plane direction, it contracted in the out of plane direction. Therefore, as the bond lengths of Fe1-Te and Fe3-Te were reduced in the plane, the bond lengths of Fe3-Te in the vertical direction were increased (Figure 7k). Biaxial strain caused a change in bond length, thereby promoting charge transfer within the monolayer (Figure 7l). When the biaxial strain increased from −5% to 5%, the spin splitting of the Fe1 3d orbital near the Fermi level became larger, as shown in Figure 7m; after the strain was applied, the spin polarization of Fe3 atoms would increase in Figure 7n, which would cause the magnetic moment to increase. In conclusion, biaxial strain-mediated FM in the FGT monolayer was closely related to the charge transfer between Fe1 and Te atoms.

Previous studies have mainly focused on biaxial strain-mediated FM in the FGT monolayers [82,85], without studying the modulation mechanism of uniaxial strain on FM. As a typical example, Zhu et el. [86] investigated the uniaxial (ε_a and ε_c) and isotropic ($\varepsilon_{iso.}$) strain modulation of the exchange coupling constant J_{ij} and T_C in FGT. As shown in Figure 8, three different magnetic configurations (FM, AFM1, and AFM2) were considered. The total energy of AFM2 was much higher than those of FM and AFM1.

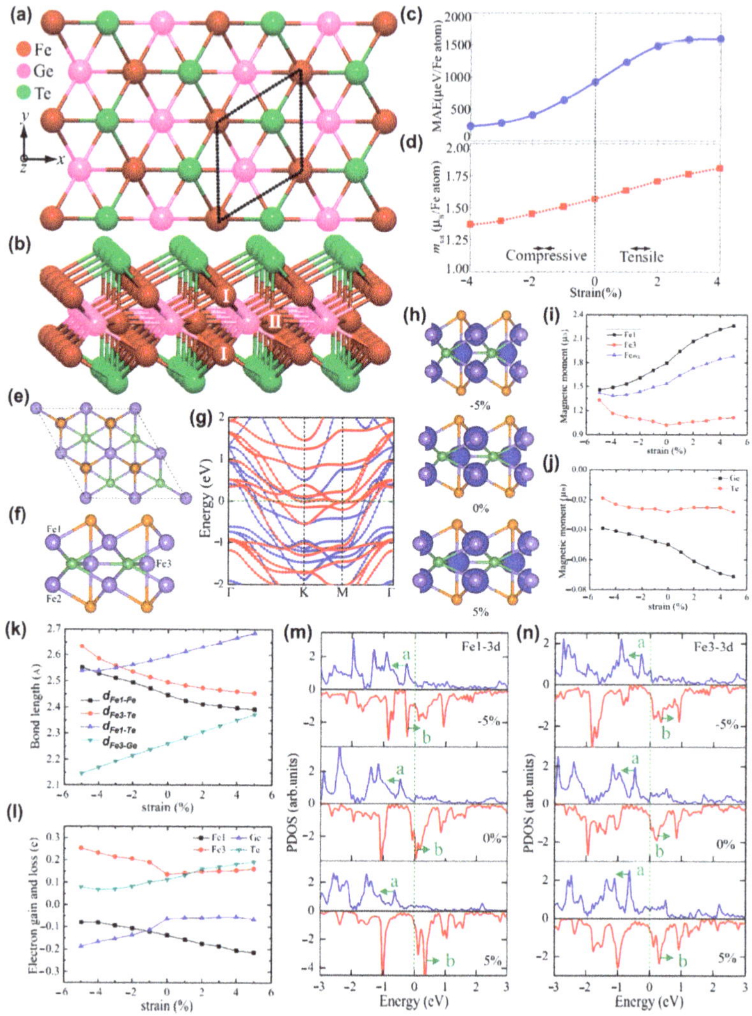

Figure 7. (**a**) Top and (**b**) side views of the atomic structure of monolayer Fe_3GeTe_2. The unit cell is enclosed by the dotted lines. Inequivalent Fe sites are numbered by I and II, respectively. Variation of (**c**) MAE and (**d**) total magnetic moment per Fe atom of single-layer Fe_3GeTe_2 under biaxial strain. (Reproduced with permission from [82]. Copyright 2016, American Physical Society). (**e**) Top and (**f**) side views of the structural models of the FGT monolayer. The purple, green, and yellow balls stand for Fe, Ge, and Te atoms, respectively. Fe sites are numbered by Fe1, Fe2, and Fe3, respectively, where Fe1 and Fe2 atoms are located at 2 equivsites, while Fe3 atom has the inequivalent site. (**g**) Spin-polarized band structures of the FGT monolayer. The Fermi level is set at zero, denoted by the olive dashed line. (**h**) Spin-density distribution of the FGT monolayer with −5, 0, and 5% strain. The isovalues are 0.02 e/Å^3. Strain dependence of magnetic moment (**i**) per Fe1 and Fe3 atoms and (**j**) per Ge and Te atoms in the FGT monolayer. Strain dependence of (**k**) the distance and the bonding length (Fe1–Fe2 distance, $d_{Fe1\text{-}Fe2}$; Fe3–Te bond length, $d_{Fe3\text{-}Te}$; Fe1−Te bond length, $d_{Fe1\text{-}Te}$; Fe3–Ge bond length, $d_{Fe3\text{-}Ge}$). (**l**) Electron transfer of Fe, Ge, and Te atoms in the FGT monolayer. PDOS (partial density of states) of (**m**) Fe1 atom and (**n**) Fe3 atom in the FGT monolayer. The enhanced spin splitting of the a and b states in PDOS. (Reproduced with permission from [85]. Copyright 2020, American Chemical Society).

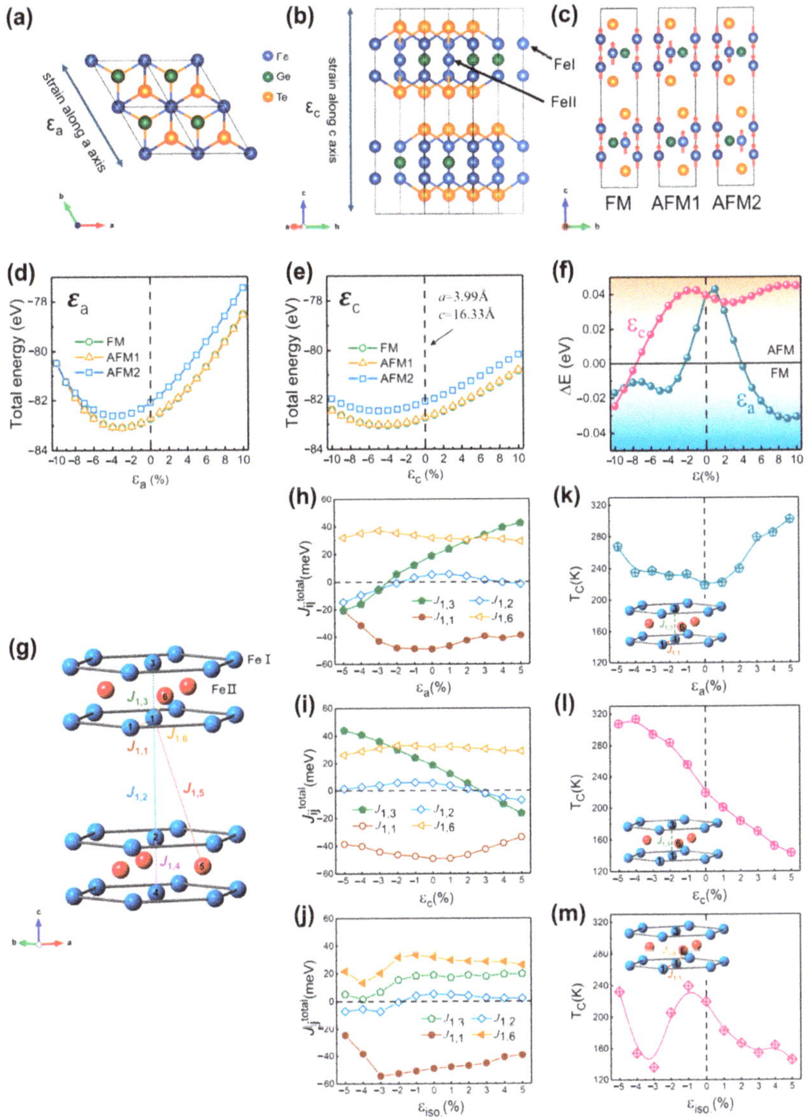

Figure 8. Schematic of the crystal structure of FGT from (**a**) top and (**b**) side view, and (**c**) the spin structure of FM, AFM1, and AFM2 configurations. (**d**,**e**) Total energy of FM, AFM1, and AFM2 configurations of FGT as a function of lattice distortion (**d**) along the a-axis and (**e**) the c-axis. (**f**) Total energy difference ($\Delta E = E_{FM} - E_{AFM}$) under c-axis and a-axis strain. The background indicates the FM- or AFM-stable region. (**g**) The schematic picture of Fe–Fe exchange interactions in FGT, where only the Fe atoms are displayed. (**h**–**j**) Total isotropic exchange coupling parameters of $J_{1,1}$, $J_{1,2}$, $J_{1,3}$, $J_{1,6}$ in functions of three different kinds of distortion: (**h**) along the a-axis, (**i**) along the c-axis and (**j**) isotropically. (**k**–**m**) Corresponding variation of T_C under above strains. The insets show the $J_{i,j}$ that play a main role in the variation of T_C. (Reproduced with permission from [86]. Copyright 2018, Elsevier).

The system exhibited a phase transition from AFM to FM, similar to the modulation result of biaxial strain [59]. Interestingly, when the strain (ε_a) along the a-axis was applied, the phase transition occurred at a tensile of ~4% or compressive of ~2%; when the strain (ε_c) along the c-axis strain was applied, it occurred at a compressive of ~8–10%. Furthermore, the strain-mediated variation of J_{ij}^{total} is shown in Figure 8g–i. T_C could be estimated by the following equation [86,98,99]:

$$T_C = \frac{2}{3K_B} J_{max} \tag{2}$$

Note that J_{max} is the maximal eigenvalue of the matrix consisting of the exchange coupling between different atoms [86,109]. After applying the uniaxial strain, the T_C could be raised to room temperature (Figure 8k–m). When isotropic strain is applied, the T_C exhibited complex nonlinear changes and failed to reach room temperature (Figure 8m). Their results demonstrated that applying uniaxial strain was an effective way to elevate the T_C.

Regarding another interesting compound, Fe_5GeTe_2 [110,111], recent experiments have found that it has a higher T_C. However, little is known about the electronic and magnetic properties of its monolayer. Joe et al. [83] found that the magnetism of bulk and monolayer metallic Fe_5GeTe_2 originated from Fe d orbital. Moreover, biaxial strain could also enhance the Fe magnetic moment from 1.65 μ_B to 2.66 μ_B.

3. Introducing Strain in 2D van der Waals Materials

3.1. Wrinkle-Induced Strain

The previously discovered strain-mediated FM phenomena were all in 2D materials without intrinsic long-range magnetic order [27,28,41]. Very recently, Seidel et al. [42] found a clear dependence of T_C on the strain state and the thickness of $Cr_2Ge_2T_6$ (CGT). The layered CGT lattice with intrinsic magnetism was shown in Figure 9a. A stripe domain structure was observed with MFM (Figure 9b). After buckling, the wrinkled area had a higher T_C than the flat area on (Figure 9c). DFT calculations indicated that the strain could elevate the T_C in monolayer and bilayer $Cr_2Ge_2T_6$ (Figure 9d–g). In addition, the strain of three different wrinkles in Figure 9h was confirmed by COMSOL simulations. To the best of our knowledge, this was the first time that the strain distribution in CGT wrinkles was mapped with the COMSOL simulation. Furthermore, the magnetic signals could be observed at the wrinkled regions at RT through MFM.

3.2. Bending or Pre-Stretching Flexible Substrates

Similar to pre-stretched flexible substrates, including Gel-Film [30], polydimethylsiloxane (PDMS) [30,112–117], polyethyleneterephthalate (PET) [118], polyimide (PI) [47,119,120], and polyvinyl alcohol (PVA) [121,122], strain can also be introduced into 2D materials by bending flexible substrates. As a typical example, Yan et al. [121] designed a novel polymer-buried strategy to apply tensile strain on Fe_3GeTe_2 (FGT) nanoflakes (Figure 10a). Firstly, PVA was spin-coated onto pre-stretched mechanical exfoliated FTG nanosheets. Then, the PET sheet was attached to the surface of the PVA film and peeled off with tweezers. The complex film (FGT/PVA/PET) was placed into a non-magnetic plastic tube and then the tensile strain was applied on the FGT nanosheets through a three-point fixing device. As the applied strain gradually increased, the sample transitioned from its original PM state to FM states, as shown in Figure 10b,c. When the strain reached 4.7%, the hysteresis loop evolved from soft magnet to hard magnet; however, when the strain reached 7.0%, the hysteresis loop displayed an opposite evolution trend. Moreover, the M_s and H_c were very close to those found for the case of tensile strain. Unlike zero-strain samples, the sample with a 3.4% tensile strain always exhibited centrosymmetry and no exchange bias (Figure 10d,e). The hidden AFM state and interface–exchange interaction could be revealed by controlling the strain. The frequency shift was inverted after relaxing the strain, indicating the transition from FM state to PM state (Figure 10f,g). As the strain increased, the T_C was elevated above RT in Figure 10h,i. In short, strain engineering is an efficient way to increase T_C compared to other methods (Figure 10j).

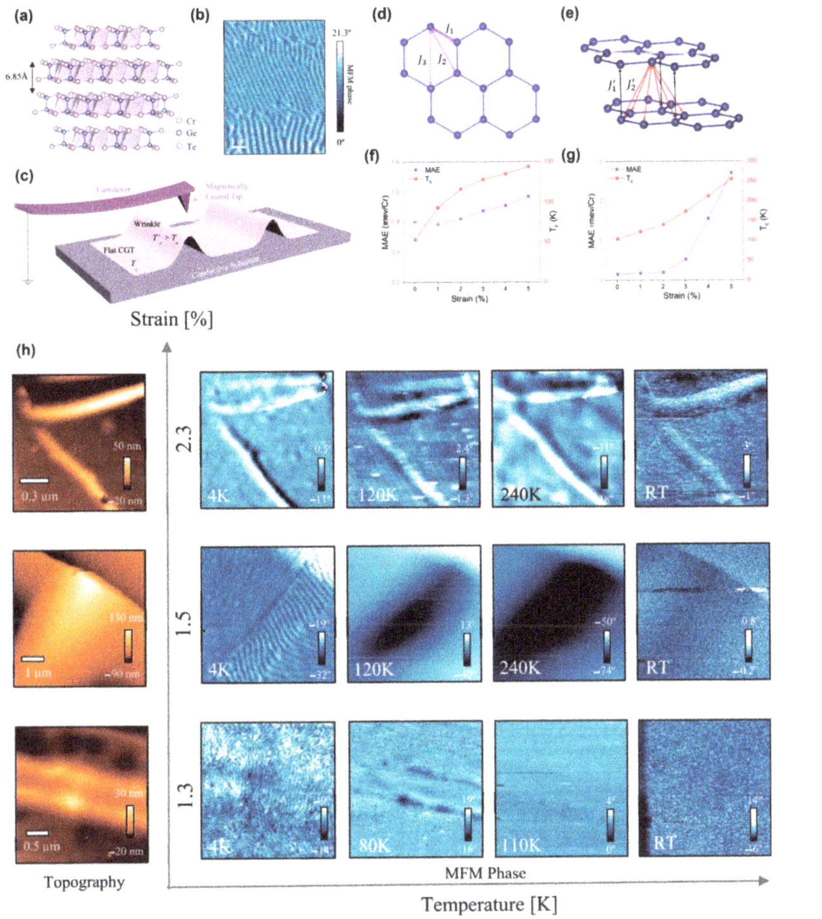

Figure 9. Curved nanostructures in *vdW* $Cr_2Ge_2Te_6$: (**a**) CGT crystal lattice showing the *vdW* layered structure and interlayer distance. (**b**) Typical magnetic stripe domain structure at 4 K seen in MFM measurements. (**c**) Schematic showing wrinkles in layered CGT and magnetically coated tip for MFM measurements. DFT predicted trend of magnetism and transition temperature with strain percentage. (**d**) Intralayer Cr the nearest-neighbor (J_1), the second-nearest−neighbor (J_2), and the third-nearest-neighbor (J_3) exchange couplings. (**e**) Interlayer Cr the nearest−neighbor ($J_{1'}$) and the second-nearest-neighbor ($J_{2'}$) exchange couplings in bilayer $Cr_2Ge_2Te_6$. The calculated MAE per Cr atom and T_C as functions of strain for (**f**) monolayer and (**g**) bilayer $Cr_2Ge_2Te_6$. (**h**) Temperature-dependent MFM examination of curved wrinkles with increasing strain: Left, Topography of three different strained wrinkles. Right, MFM image series depicting enhanced magnetic signal at the wrinkles depending on specific strain state up to RT. Magnetic signals in wrinkles exhibiting strain of 1.3% disappear above 100 K, while wrinkles with 2.3% strain exhibit clear MFM phase signals up to RT. Diagonal periodic lines in the figures are a result of instrument noise and are not part of the magnetic signal (1.3% strain at 110 K and 2.3% strain at 120 K). (Reproduced with permission from [42]. Copyright 2022, American Chemical Society).

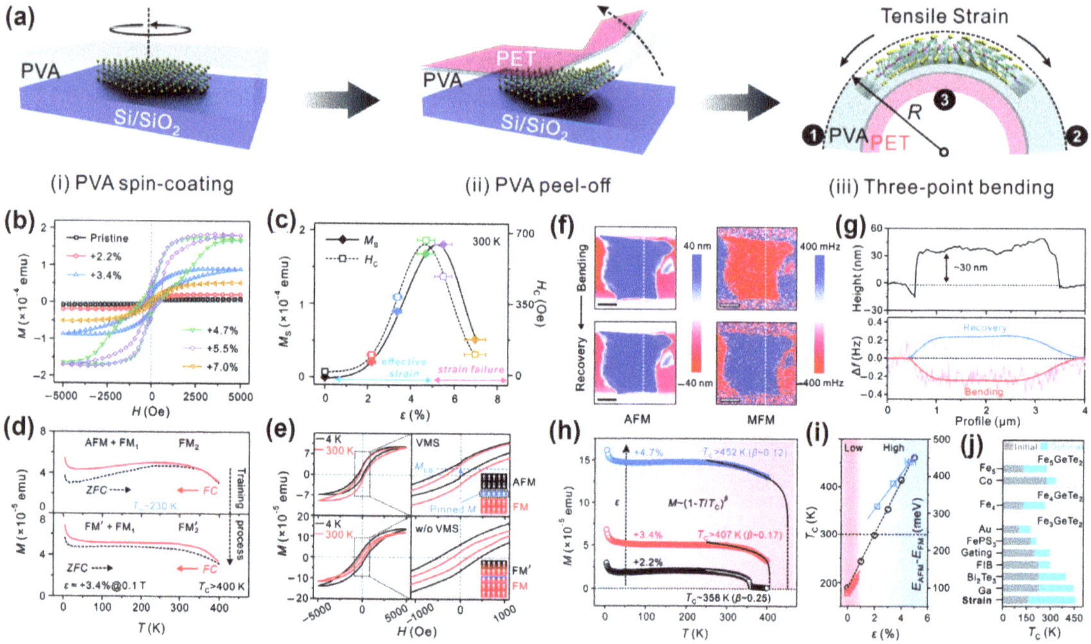

Figure 10. (**a**) Schematic illustration of the transfer and bending process of FGT nanoflakes on flexible PVA/PET substrates. (**b**) *M–H* curves at under different strain values. (**c**) Extracted M_S and H_C as a function of tensile strain. (**d**) ZFC and FC curves before and after the training procedure. (**e**) *M–H* curves and anomalous exchange-bias effect. Insets are the corresponding schematic illustrations. (**f**) AFM and MFM images of a single nanoflake in the bending and recovery states. All scale bars are 1 mm. (**g**) Extracted height and Df along the white dashed lines marked in (**f**). (**h**) *M-T* and critical fittings under different strains. (**i**) Strain-tuned T_C phase diagram. The blue square symbols are the results of this current work and others are collected from the literature. (**j**) Comparison of optimal T_C values using different methods based on the FGT system. From top to bottom: Fe_5 [111] and Co [123] represent pure and Co-doped Fe_5GeTe_2, respectively; Fe_4 refers to Fe_4GeTe_2 [124]; Au [125], $FePS_3$ [37] and Bi_2Te_3 [34] denote the corresponding heterojunctions with Fe_3GeTe_2. Gating means the electrostatically gated Fe_3GeTe_2 [126], FIB [127] and Ga [128] are focused ion beam-treated and Ga ion-implanted Fe_3GeTe_2, respectively. (Reproduced with permission from [121]. Copyright 2021, The Royal Society of Chemistry).

As another typical example, Miao et al. [47] found that when 0.32% uniaxial tensile strain was applied to the FGT nano sheet, its H_c increased by more than 150%. In order to apply strain, they directly exfoliated the FGT nanosheets onto the PI film. By controlling the distance of the needle tip pushed at the center of the substrate, the PI film was bent to apply different uniaxial tensile strains to the FGT sample. The difference of magnetic anisotropy energy was attributed to the strain-mediated FM. More importantly, they realized a magnetization reversal with the limited strain. Similarly, Xu et al. [48] also observed a reversible phase transition from AFM to FM in a CrSBr nanoflake at cryogenic temperature. Their strain equipment consisted of three piezoelectric actuators glued to a titanium flexure element. By cleaving a silicon substrate to form a micrometer-scale gap, the sample was suspended at the gaps. Furthermore, a piezo voltage was used to continuously apply strain reversibly to the CrSBr flake.

3.3. Lattice Mismatch

High-quality, single-crystalline Fe_4GeTe_2 thin films [43] were grown on sapphire substrate by molecular beam epitaxy (MBE). As shown in Figure 11a, the thin films exhibited a rhombohedral structure. During the MBE epitaxy preparation, due to a lattice mismatch of about 20% between the Fe_4GeTe_2 sample and the sapphire substrate, the lattice rotated 30° to form a perfect single crystal.

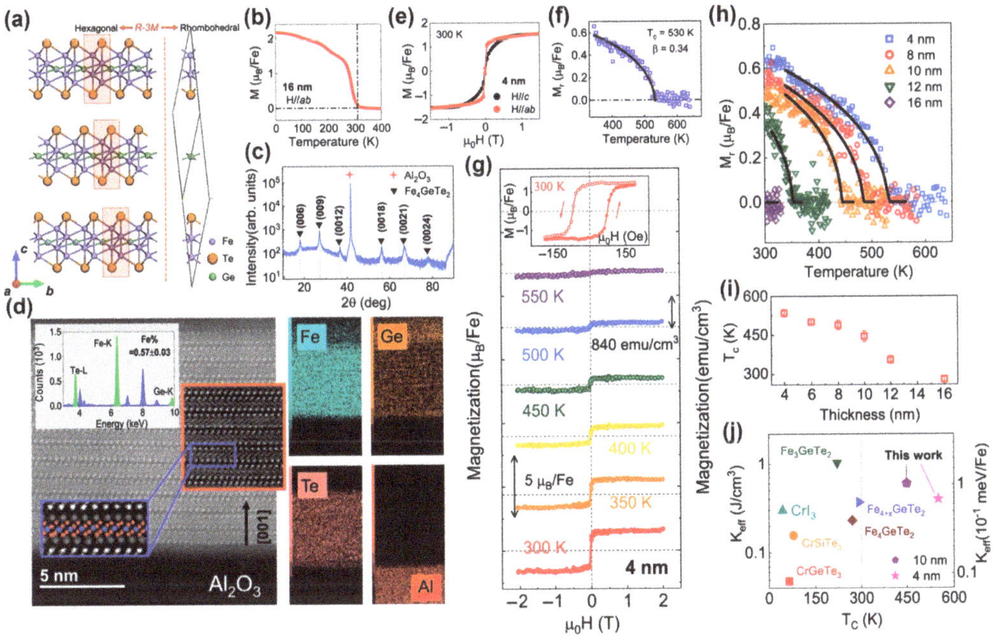

Figure 11. Crystal structure of the Fe_4GeTe_2 and its thickness-dependent magnetism. (**a**) Schematics of the crystal structure of Fe_4GeTe_2 stacked in ABC configuration (**left**), and its rhombohedral structure unit (**right**). (**b**) Temperature dependence of the magnetization for 16 nm Fe_4GeTe_2. (**c**) XRD scan of the Fe_4GeTe_2 film. (**d**) A typical HRTEM image of Fe_4GeTe_2 films; the color squares show the high-pass filtered images of the vdW structure. Up Inset: The EDX result verifies the 4:1:2 Fe: Ge: Te stoichiometric composition with the uniform element distribution map (right). I *M–H* curves of 4 nm Fe_4GeTe_2 at 300 K. (**e**) Room-temperature magnetic hysteresis loops. (**f**) In-plane M_r–T curve. (**g**) Detailed magnetic field-dependent magnetization of 4 nm Fe_4GeTe_2 at various temperatures for $H_{//ab}$. Inset: zoom-in hysteresis loop at 300 K. (**h**) M_r-T curves. (**i**) *Tc* for Fe_4GeTe_2 thin films with different thicknesses. (**j**) Effective magnetic an isotropy energy K_{eff} and curie temperature *Tc* for our samples and previous vdW ferromagnets [77,124,129–131]. (Reproduced with permission from [43]. Copyright 2023, Springer Nature).

However, there was a tensile strain of about 2% inside the Fe_4GeTe_2 film. The in-plane *M-T* curve of the 16 nm FGT film in Figure 11b indicated that its T_C was close to RT. The results of XRD (Figure 11c) and HRTEM (Figure 11d) indicated that the FGT film had a perfect single crystal, and the atomic ratio of the Fe:Ge:Te element was about 4:1:2. At 300 K, a 4 nm film exhibited robust in-plane magnetic anisotropy, as shown in Figure 11e. By fitting the M_r-T curve in Figure 11f, the T_C could be deduced as high as 530 K. Furthermore, in Figure 11, it was confirmed through *M-H* loops that the few layered FGT sample exhibited high-temperature FM. In addition, the thickness dependence of T_C showed that when the thickness was thinner, its T_C was higher, as shown in Figure 11h,i. Notably, MBE-prepared FGT had the higher T_C and maintains a relatively high-magnetic anisotropy (Figure 11j).

Through DFT calculations, it was found that a 2% tensile strain was not the reason for the enhanced T_C.

3.4. Electrostatic Force

Considering the abnormally large mechanical response of $Cr_2Ge_2Te_6$ (CGT) thin films caused by strong magnetostriction at T_C, it was expected that this reverse effect could achieve strain-controlled T_C. As shown in Figure 12a–c, the electrostatic strain-tuning T_C was realized in a suspended $Cr_2Ge_2Te_6/WSe_2$ heterostructure [44]. After applying a gate voltage V_g, there was electrostatic force (F_{el}) in Figure 12c between the heterostructure and the bottom silicon substrate, resulting in strain in $Cr_2Ge_2Te_6$ thin films. The strain of the CGT layer can be estimated using the following formula:

$$\begin{aligned}\Delta\epsilon &= \epsilon_{el}(V_g) + \epsilon_b(V_g) \\ &\approx \left[\frac{2}{3}\left(\frac{\varepsilon_0 r}{8{g_0}^2 n(T_0)}\right)^2\right] V_g^4 + \left[\frac{\varepsilon_0 r \Delta z(1-v)}{4{g_0}^2 n(T_0)}\right] V_g^2,\end{aligned} \quad (3)$$

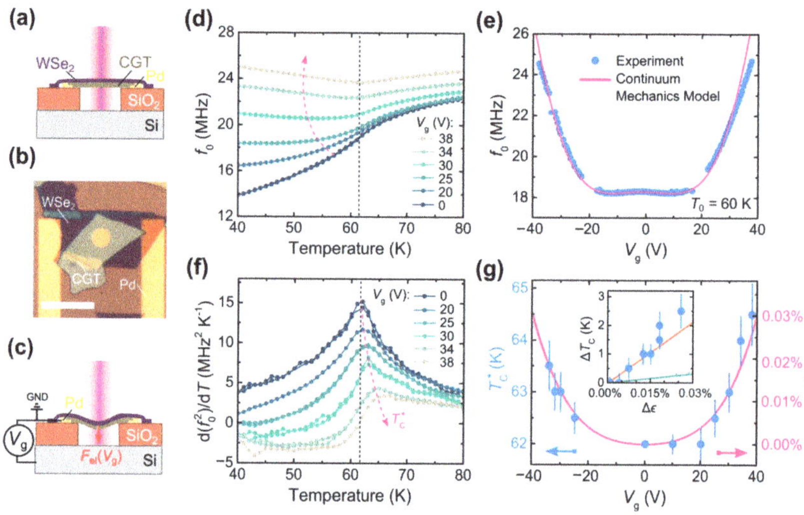

Figure 12. Strain-mediated T_C in a suspended CGT/WSe_2 heterostructure. (**a**) The schematic cross-section of the suspended CGT/WSe_2 heterostructure membrane. (**b**) The optical image of the heterostructure. Scale bar: 12 μm. (**c**) The schematics of the electrostatic strain-tuning principle. (**d**) Measured resonance frequencies f_0 of the heterostructure membrane as a function of temperature for different gate voltages V_g. (**e**) Filled blue circles—the measured resonance frequency as a function of V_g at 60 K. (**f**) The temperature derivative of ${f_0}^2$ as a function of temperature. (**g**) Solid magenta line—the estimate of electrostatically induced strain $\Delta\epsilon$ as a function of V_g. The shaded magenta region shows the uncertainty in $\Delta\epsilon$. The inset shows as a function of added voltage-induced strain $\Delta\epsilon$ in addition to calculations from Li and Yang [73] (solid green line) and from Dong et al. [132] (solid orange line). The vertical error bars in T_C were estimated from determining the peak position in (**e**) within 2% accuracy in the measured maximum. (Reproduced with permission from [44]. Copyright 2022, American Chemical Society).

Note that $\Delta\epsilon$ is the total strain in the CGT layer; ϵ_{el} is the electrostatic pulling strain; ϵ_b is bending strain; ε_0 is the dielectric constant of vacuum; V_g is the gate voltage; r is the membrane radius; v is the Poisson's ratio; $n(T_0)$ is total tension in suspended heterostructures at $T_0 = 60$ K; and g_0 is the separation between the heterostructure and the bottom Si substrate. Interestingly, when the electrostatic force induced a strain of 0.026%, the T_C of the suspended $Cr_2Ge_2Te_6$ /WSe_2 heterostructure was increased by about 2.6 K, as shown in Figure 12d–g.

3.5. Field-Cooling

Due to the fact that the lattice parameters of CGT materials at 5 K are greater than those at 270 K, it would cause in-plane expansion during cooling [133,134]. Phatak et al. [45] found that field-cooling could cause the material expansion in CGT flakes, leading to in-plane strain. Furthermore, they directly observed strain-induced evolution of the magnetic domain structure by cryogenic Lorentz transmission electron microscopy (LTEM), which was closely related to the magnetoelastic coupling between strain and magnetization. This work revealed how to directly measure magnetic domain structures at the nanoscale.

4. Conclusions and Outlook

In this review, we have summarized the recent progress of strain-mediated intrinsic FM in 2D van der Waals materials with long-range order. First, we introduce how to explain the strain-mediated intrinsic FM on Cr-based and Fe-based 2D van der Waals materials with long-range FM order through ab initio Density functional theory (DFT), and how to calculate magnetic anisotropy energy (MAE) and T_C from the interlayer exchange coupling. Subsequently, we focus on numerous attempts to apply strain to 2D materials in experiments, including wrinkle-induced strain, flexible substrate bending or stretching, lattice mismatch, electrostatic force and field-cooling. However, research in this field is still in its early stages and there are many challenges that must be overcome.

By selecting substrates with different thermal expansion coefficients, including x-quartz [135–138], sapphire [27,28,139], and SiO_2 [114]), uniaxial [135–138] or biaxial strain [27,28,114,139,140] could be introduced into 2D materials. The introduction of biaxial strain in this way was limited to the study of 2D materials without intrinsic long-range magnetic order, such as MoS_2 [28] and ReS_2 [27]. So far, this strategy has not been applied to the research on 2D intrinsic van der Waals materials. Very few equipment [47,48,121] could apply strain to 2D materials, but less could be coupled with magnetic testing equipment [41,141]. Especially, it is very challenging that one could accurately determine the strain while measuring the magnetic properties and T_C in different regions.

Recently, some emerging magnetic imaging technologies, such as magneto-optical Kerr effect (MOKE) [7,8,40,142–144], magnetic circular dichroism (MCD) [62,126,145–149], photoemission electron microscopy (PPEM) [150], scanning transmission X-ray microscopy (STXM) [151], Lorentz transmission electron microscopy (LTEM) [150–152], spin-polarized scanning tunneling microscopy (SP-TEM) [66,153], MFM [41,154], scanning SQUID [155–159], and scanning nitrogen-vacancy center microscopy (SNVM) [160–164], have been used to study 2D FM materials. However, these studies have not yet been combined with strain and, currently, it is almost impossible to directly analyze the relationship between strain and FM in 2D intrinsic materials. More importantly, strengthening the guideline of strain-mediated FM will promote the development of spintronics and straintronics.

Author Contributions: Writing—original draft preparation, H.R.; writing—review and editing, H.R. and G.X.; supervision, G.X. All authors have read and agreed to the published version of the manuscript.

Funding: This research was funded by National Natural Science Foundation of China (Grant No. 52172272), Shandong Province Natural Science Foundation (Grant No. ZR202103040767) and Doctoral Scientific Research Foundation of Liaocheng University (Grant No. 318052054).

Institutional Review Board Statement: Not applicable.

Informed Consent Statement: Not applicable.

Data Availability Statement: Not applicable.

Conflicts of Interest: The authors declare no conflict of interest.

Abbreviations

2D	two-dimensional
3D	three-dimensional
AFM	antiferromagnetism
BKT	Berezinskii-Kosterlitz-Thouless
CDW	charge density wave
CGT	$Cr_2Ge_2Te_6$
CPT	$CrPbTe_3$
DFT	density functional theory
DM	Dzialoshinskii-Moriya
DOS	density of states
FGT	Fe_3GeTe_2
FM	ferromagnetism
GGA	the generalized gradient approach
HRTEM	high resolution transmission electron microscopy
LDA	linear density approach
LRO	long-range order
LSDA	local spin density approximation
LTEM	lorentz transmission electron microscopy
MAE	magnetic anisotropy energy
MBE	molecular beam epitaxy
MC	Monte Carlo
MCD	magnetic circular dichroism
MFM	magnetic force microscopy
MOKE	magneto-optical Kerr effect
NM	nonmagnetism
PAD	polymer assisted deposition
PDMS	polydimethylsiloxane
PEEM	photoemission electron microscopy
PET	polyethyleneterephthalate
PI	polyimide
PRA	the random phase approximation
PVA	polyvinyl alcohol
QMC	Quantum Monte Carlo
RT	room temperature
SEM	scanning electron microscopy
SNVM	scanning nitrogen-vacancy center microscopy
SP-TEM	spin-polarized scanning tunneling microscopy
SQUID	superconducting quantum interference device magnetometry
SOC	spin-orbit coupling
SSR	solid-state reaction
STM	scanning tunneling microscopy
STXM	scanning transmission X-ray microscopy
XRD	X-ray diffraction
F_{el}	electrostatic force
T_C	curie temperature
J	the exchange coupling constant
K	magnetic anisotropy
ΔE	total energy difference

References

1. Mermin, N.D.; Wagner, H. Absence of ferromagnetism or antiferromagnetism in one- or two-dimensional isotropic heisenberg models. *Phys. Rev. Lett.* **1966**, *17*, 1133–1136. [CrossRef]
2. Hohenberg, P.C. Existence of long-range order in one and two dimensions. *Phys. Rev.* **1967**, *158*, 383–386. [CrossRef]
3. Meiklejohn, W.H.; Bean, C.P. New magnetic anisotropy. *Phys. Rev.* **1956**, *102*, 1413–1414. [CrossRef]
4. Onsager, L. Crystal statistics. I. a two-dimensional model with an order-disorder transition. *Phys. Rev.* **1944**, *65*, 117–149. [CrossRef]

5. Gibertini, M.; Koperski, M.; Morpurgo, A.F.; Novoselov, K.S. Magnetic 2D materials and heterostructures. *Nat. Nanotechnol.* **2019**, *14*, 408–419. [CrossRef] [PubMed]
6. Wang, Q.H.; Bedoya-Pinto, A.M.; Dismukes, A.H.; Hamo, A.; Jenkins, S.; Koperski, M.; Liu, Y.; Sun, Q.C.; Telford, E.J.; Kim, H.H.; et al. The magnetic genome of two-dimensional van der Waals materials. *ACS Nano* **2022**, *16*, 6960–7079. [CrossRef]
7. Gong, C.; Li, L.; Li, Z.L.; Ji, H.W.; Stern, A.; Xia, Y.; Cao, T.; Bao, W.; Wang, C.Z.; Wang, Y.A.; et al. Discovery of intrinsic ferromagnetism in two-dimensional van der Waals crystals. *Nature* **2017**, *546*, 265–269. [CrossRef]
8. Huang, B.; Clark, G.; Navarro-Moratalla, E.; Klein, D.R.; Cheng, R.; Seyler, K.L.; Zhong, D.; Schmidgall, E.; McGuire, M.A.; Cobden, D.H.; et al. Layer-dependent ferromagnetism in a van der Waals crystal down to the monolayer limit. *Nature* **2017**, *546*, 270–273. [CrossRef]
9. Bissett, M.A.; Tsuji, M.; Ago, H. Strain engineering the properties of graphene and other two-dimensional crystals. *Phys. Chem. Chem. Phys.* **2014**, *16*, 11124–11138. [CrossRef]
10. Roldán, R.; Castellanos-Gomez, A.; Cappelluti, E.; Guinea, F. Strain engineering in semiconducting two-dimensional crystals. *J. Phys. Condens. Matter* **2019**, *27*, 313201. [CrossRef]
11. Luo, M.C.; Guo, S.J. Strain-controlled electrocatalysis on multimetallic nanomaterials. *Nat. Rev. Mater.* **2019**, *2*, 17059. [CrossRef]
12. Bukharaev, A.A.; Zvezdin, K.; Pyatakov, A.P.; Fetisov, Y.K. Straintronics: A new trend in micro- and nanoelectronics and materials science. *Phys.-Usp.* **2018**, *61*, 1175–1212. [CrossRef]
13. Deng, S.K.; Sumant, A.V.; Berry, V. Strain engineering in two-dimensional nanomaterials beyond graphene. *Nano Today* **2018**, *22*, 14–35. [CrossRef]
14. Dai, Z.H.; Liu, L.Q.; Zhang, Z. Strain engineering of 2D materials: Issues and opportunities at the interface. *Adv. Mater.* **2019**, *31*, 1805417. [CrossRef] [PubMed]
15. Xia, Z.H.; Guo, S.J. Strain engineering of metal-based nanomaterials for energy electrocatalysis. *Chem. Soc. Rev.* **2019**, *48*, 3265–3278. [CrossRef] [PubMed]
16. Sun, Y.F.; Liu, K. Strain engineering in functional 2-dimensional materials. *J. Appl. Phys.* **2019**, *125*, 082402. [CrossRef]
17. Peng, Z.W.; Chen, X.L.; Fan, Y.L.; Srolovitz, D.J.; Lei, D.Y. Strain engineering of 2D semiconductors and graphene: From strain fields to band-structure tuning and photonic applications. *Light-Sci. Appl.* **2020**, *9*, 190. [CrossRef]
18. Yan, Y.L.; Ding, S.; Wu, X.N.; Zhu, J.; Feng, D.M.; Yang, X.D.; Li, F.F. Tuning the physical properties of ultrathin transition-metal dichalcogenides via strain engineering. *RSC Adv.* **2020**, *10*, 39455–39467. [CrossRef]
19. Miao, F.; Liang, S.J.; Cheng, B. Straintronics with van der Waals materials. *NPJ Quantum Mater.* **2021**, *6*, 59. [CrossRef]
20. Ren, H.T.; Xiang, G. Morphology-dependent room-temperature ferromagnetism in undoped ZnO nanostructures. *Nanomaterials* **2021**, *11*, 3199. [CrossRef]
21. Yang, S.X.; Chen, Y.J.; Jiang, C.B. Strain engineering of two-dimensional materials: Methods, properties, and applications. *InfoMat* **2021**, *3*, 397–420. [CrossRef]
22. Xu, X.H.; Liang, T.; Kong, D.B.; Wang, B.; Zhi, L.J. Strain engineering of two-dimensional materials for advanced electrocatalysts. *Mater. Today Nano* **2021**, *41*, 100111. [CrossRef]
23. Yang, X.B.; Wang, Y.Y.; Tong, X.L.; Yang, N.J. Strain engineering in electrocatalysts: Fundamentals, progress, and perspectives. *Adv. Energy Mater.* **2022**, *12*, 2102261. [CrossRef]
24. Qi, Y.P.; Sadi, M.A.; Hu, D.; Zheng, M.; Wu, Z.P.; Jiang, Y.C.; Chen, Y.P. Recent progress in strain engineering on van der Waals 2D materials: Tunable electrical, electrochemical, magnetic, and optical properties. *Adv. Mater.* **2023**, *35*, 2205714. [CrossRef] [PubMed]
25. Ren, H.T.; Xiang, G. Strain-modulated magnetism in MoS_2. *Nanomaterials* **2022**, *12*, 1929. [CrossRef] [PubMed]
26. Ren, H.T.; Xiang, G. Recent progress in research on ferromagnetic rhenium disulfide. *Nanomaterials* **2022**, *19*, 3451. [CrossRef]
27. Ren, H.T.; Xiang, G.; Lu, J.T.; Zhang, X.; Zhang, L. Biaxial strain-mediated room temperature ferromagnetism of ReS_2 web buckles. *Adv. Electron. Mater.* **2019**, *5*, 1900814. [CrossRef]
28. Ren, H.T.; Zhang, L.; Xiang, G. Web buckle-mediated room-temperature ferromagnetism in strained MoS_2 thin films. *Appl. Phys. Lett.* **2020**, *116*, 012401. [CrossRef]
29. Novoselov, K.S.; Geim, A.K.; Morozov, S.V.; Jiang, D.; Zhang, Y.; Dubonos, S.V.; Grigorieva, I.V.; Firsov, A.A. Electric field effect in atomically thin carbon films. *Science* **2004**, *306*, 666. [CrossRef]
30. Castellanos-Gomez, A.; Roldan, R.; Cappelluti, E.; Buscema, M.; Guinea, F.; van der Zant, H.S.J.; Steele, G.A. Local strain engineering in atomically rhin MoS_2. *Nano Lett.* **2013**, *13*, 5361–5366. [CrossRef]
31. Lu, P.; Wu, X.J.; Guo, W.L.; Zeng, X.C. Strain-dependent electronic and magnetic properties of MoS_2 monolayer, bilayer, nanoribbons and nanotubes. *Phys. Chem. Chem. Phys.* **2012**, *14*, 13035–13040. [CrossRef] [PubMed]
32. Conley, H.J.; Wang, B.; Ziegler, J.I.; Haglund, R.F.; Pantelides, S.T.; Bolotin, K.I. Bandgap engineering of strained monolayer and bilayer MoS_2. *Nano Lett.* **2013**, *13*, 3626–3630. [CrossRef] [PubMed]
33. Bertolazzi, S.; Brivio, J.; Kis, A. Stretching and breaking of ultrathin MoS_2. *ACS Nano* **2011**, *5*, 9703–9709. [CrossRef]
34. Wang, H.Y.; Liu, Y.J.; Wu, P.C.; Hou, W.J.; Jiang, Y.H.; Li, X.H.; Pandey, C.; Chen, D.D.; Yang, Q.; Wang, H.T.; et al. Above room-temperature ferromagnetism in wafer-scale two-dimensional van der Waals Fe_3GeTe_2 tailored by a topological insulator. *ACS Nano* **2020**, *14*, 10045–10053. [CrossRef] [PubMed]

35. Chen, X.H.; Wang, H.T.; Liu, H.J.; Wang, C.; Wei, G.S.; Fang, C.; Wang, H.C.; Geng, C.Y.; Liu, S.J.; Li, P.Y.; et al. Generation and control of Terahertz spin currents in topology-induced 2D ferromagnetic Fe_3GeTe_2 | Bi_2Te_3 heterostructures. *Adv. Mater.* **2022**, *34*, 2106172. [CrossRef]
36. Dolui, K.; Petrovic, M.D.; Zollner, K.; Plechac, P.; Fabian, J.; Nikolic, B.K. Proximity spin-orbit torque on a two-dimensional magnet within van der Waals heterostructure: Current-driven antiferromagnet-to-ferromagnet reversible nonequilibrium phase transition in Bilayer CrI_3. *Nano Lett.* **2020**, *20*, 2288–2295. [CrossRef]
37. Zhang, L.M.; Huang, X.Y.; Dai, H.W.; Wang, M.S.; Cheng, H.; Tong, L.; Li, Z.; Han, X.T.; Wang, X.; Ye, L.; et al. Proximity-coupling-induced significant enhancement of coercive field and curie temperature in 2D van der Waals heterostructures. *Adv. Mater.* **2020**, *34*, 2002032. [CrossRef]
38. Wang, Z.Y.; Tang, C.; Sachs, R.; Barlas, Y.; Shi, J. Proximity-induced ferromagnetism in graphene revealed by the anomalous hall effect. *Phys. Rev. Lett.* **2015**, *114*, 016603. [CrossRef]
39. Katmis, F.; Lauter, V.; Nogueira, F.S.; Assaf, B.A.; Jamer, M.E.; Wei, P.; Satpati, B.; Freeland, J.W.; Eremin, I.; Heiman, D.; et al. A high-temperature ferromagnetic topological insulating phase by proximity coupling. *Nature* **2016**, *533*, 513–516. [CrossRef]
40. Ciorciaro, L.; Kroner, M.; Watanabe, K.; Taniguchi, T.; Imamoglu, A. Observation of magnetic proximity effect using resonant optical spectroscopy of an electrically tunable $MoSe_2/CrBr_3$ heterostructure. *Phys. Rev. Lett.* **2020**, *124*, 197401. [CrossRef]
41. Yang, S.X.; Wang, C.; Sahin, H.; Chen, H.; Li, Y.; Li, S.S.; Suslu, A.; Peeters, F.M.; Liu, Q.; Li, J.B.; et al. Tuning the optical, magnetic, and electrical properties of $ReSe_2$ by nanoscale strain engineering. *Nano Lett.* **2015**, *15*, 1660–1666. [CrossRef] [PubMed]
42. O'Neill, A.; Rahman, S.; Zhang, Z.; Schoenherr, P.; Yildirim, T.J.; Gu, B.; Su, G.; Lu, Y.R.; Seidel, J. Enhanced room temperature ferromagnetism in highly strained 2D semiconductor $Cr_2Ge_2Te_6$. *ACS Nano* **2023**, *17*, 735–742. [CrossRef]
43. Wang, H.T.; Lu, H.C.; Guo, Z.X.; Li, A.; Wu, P.C.; Li, J.; Xie, W.R.; Sun, Z.M.; Li, P.; Damas, H.; et al. Interfacial engineering of ferromagnetism in wafer-scale van der Waals Fe_4GeTe_2 far above room temperature. *Nat. Commun.* **2023**, *14*, 2483. [CrossRef] [PubMed]
44. Šiškins, M.; Kurdi, S.; Lee, M.; Slotboom, B.J.M.; Xing, W.Y.; l Mañas-Valero, S.; Coronado, E.; Jia, S.; Han, W.; Sar, T.V.D.; et al. Nanomechanical probing and strain tuning of the curie temperature in suspended $Cr_2Ge_2Te_6$-based heterostructures. *NPJ 2D Mater. Appl.* **2022**, *6*, 41. [CrossRef]
45. McCray, A.R.C.; Li, Y.; Qian, E.; Li, Y.; Wang, W.; Huang, Z.J.; Ma, X.M.; Liu, Y.Z.; Chung, D.Y.; Kanatzidis, M.G.; et al. Direct observation of magnetic bubble lattices and magnetoelastic effects in van der Waals $Cr_2Ge_2Te_6$. *Adv. Funct. Mater.* **2023**, *23*, 2214203. [CrossRef]
46. Noah, A.; Zur, Y.; Fridman, N.; Singh, S.; Gutfreund, A.; Herrera, E.; Vakahi, A.; Remennik, S.; Huber, M.E.; Gazit, S.; et al. Nano-patterned magnetic edges in $CrGeTe_3$ for quasi 1-D spintronic devices. *ACS Appl. Nano Mater.* **2023**, *6*, 8627–8634. [CrossRef]
47. Yu, W.; Wang, C.; Liang, S.J.; Ma, Z.C.; Xu, K.; Liu, X.W.; Zhang, L.L.; Admasu, A.S.; Cheong, S.W.; Wang, L.Z.; et al. Strain-sensitive magnetization reversal of a van der Waals magnet. *Adv. Mater.* **2020**, *23*, 2004533.
48. Cenker, J.; Sivakumar, S.; Xie, K.C.; Miller, A.; Thijssen, P.; Liu, Z.Y.; Dismukes, A.; Fonseca, J.; Anderson, E.; Zhu, X.Y.; et al. Reversible strain-induced magnetic phase transition in a van der Waals magnet. *Nat. Nanotechnol.* **2022**, *17*, 256–261. [CrossRef]
49. Rahman, S.; Torres, J.F.; Khan, A.R.; Lu, Y.R. Recent developments in van der Waals antiferromagnetic 2D materials: Synthesis, characterization, and device implementation. *ACS Nano* **2021**, *15*, 17175–17213. [CrossRef]
50. Liu, Y.P.; Zeng, C.; Zhong, J.H.; Ding, J.N.; Wang, Z.M.; Liu, Z.W. Spintronics in two-dimensional materials. *Nan-Micro Lett.* **2020**, *12*, 93. [CrossRef]
51. Mak, K.; Shan, J.; Ralph, D.C. Probing and controlling magnetic states in 2D layered magnetic materials. *Nat. Rev. Phys.* **2019**, *1*, 646–661. [CrossRef]
52. Marchiori, E.; Ceccarelli, L.; Rossi, N.; Lorenzelli, L.; Degen, C.L.; Poggio, M. Nanoscale magnetic field imaging for 2D materials. *Nat. Rev. Phys.* **2021**, *4*, 49–60. [CrossRef]
53. Li, W.; Yang, Z.Y.; Hou, Y.L.; Gao, S. Controllable preparation and magnetism control of two-dimensional magnetic nanomaterials. *Rrog. Chem.* **2020**, *32*, 1437–1451.
54. Xing, S.C.; Zhou, J.; Zhang, X.G.; Elliott, S.; Sun, Z.M. Theory, properties and engineering of 2D magnetic materials. *Rrog. Mater. Sci.* **2023**, *132*, 101036. [CrossRef]
55. Ningrum, V.P.; Liu, B.W.; Wang, W.; Yin, Y.; Cao, Y.; Zha, C.Y.; Xie, H.G.; Jiang, X.H.; Sun, Y.; Qin, S.C.; et al. Recent advances in two-dimensional magnets: Physics and devices towards spintronic applications. *Research* **2020**, *2020*, 1768918. [CrossRef]
56. Gong, C.; Zhang, X. Two-dimensional magnetic crystals and emergent heterostructure devices. *Science* **2019**, *363*, 6428. [CrossRef] [PubMed]
57. Dai, C.Y.; He, P.; Luo, L.X.; Zhan, P.X.; Guan, B.; Zheng, J. Research progress of two-dimensional magnetic materials. *Sci. China. Mater.* **2023**, *66*, 859–876. [CrossRef]
58. Zhang, W.B.; Qu, Q.; Zhu, P.; Lam, C.H. Robust intrinsic ferromagnetism and half semiconductivity in stable two-dimensional single-layer chromium trihalides. *J. Phys. Chem. C* **2015**, *3*, 12457–12468. [CrossRef]
59. Webster, L.; Yan, J.A. Strain-tunable magnetic anisotropy in monolayer $CrCl_3$, $CrBr_3$, and CrI_3. *Phys. Rev. B* **2018**, *98*, 144411. [CrossRef]
60. Wang, Z.; Gibertini, M.; Dumcenco, D.; Taniguchi, T.; Watanabe, K.; Giannini, E.; Morpurgo, A.F. Determining the phase diagram of atomically thin layered antiferromagnet $CrCl_3$. *Nat. Nanotechnol.* **2019**, *14*, 1116–1122. [CrossRef]

61. Ahmad, A.S.; Liang, Y.C.; Dong, M.D.; Zhou, X.F.; Fang, L.M.; Xia, Y.H.; Dai, J.H.; Yan, X.Z.; Yu, X.H.; Dai, J.F.; et al. Pressure-driven switching of magnetism in layered $CrCl_3$. *Nanoscale* **2020**, *12*, 22935–22944. [CrossRef] [PubMed]
62. Bedoya-Pinto, A.; Ji, J.R.; Pandeya, A.K.; Gargiani, P.; Valvidares, M.; Sessi, P.; Taylor, J.M.; Radu, F.; Chang, K.; Parkin, S.S.P. Intrinsic 2D-XY Ferromagnetism in a van der Waals monolayer. *Science* **2021**, *374*, 616–620. [CrossRef] [PubMed]
63. Dupont, M.; Kvashnin, Y.O.; Shiranzaei, M.; Fransson, J.; Laflorencie, N.; Kantian, A. Monolayer $CrCl_3$ as an ideal test bed for the universality classes of 2D magnetism. *Phys. Rev. Lett.* **2021**, *127*, 037204. [CrossRef]
64. Ebrahimian, A.; Dyrdał, A.; Qaiumzadeh, A. Control of magnetic states and spin interactions in bilayer $CrCl_3$ with strain and electric fields: An ab initio study. *Sci. Rep.* **2023**, *13*, 5336. [CrossRef] [PubMed]
65. Liu, J.; Mo, P.H.; Shi, M.C.; Gao, D.; Jiwu Lu, J.W. Multi-scale analysis of strain-dependent magnetocrystalline anisotropy and strain-induced villari and nagaoka-honda effects in a two-dimensional ferromagnetic chromium tri-iodide monolayer. *J. Appl. Phys.* **2018**, *124*, 044303. [CrossRef]
66. Sivadas, N.; Okamoto, S.; Xu, X.D.; Fennie, C.J.; Xiao, D. Stacking-dependent magnetism in bilayer CrI_3. *Nano Lett.* **2018**, *18*, 7658–7664. [CrossRef]
67. Li, T.X.; Jiang, S.W.; Sivadas, N.; Wang, Z.F.; Xu, Y.; Weber, D.; Goldberger, J.E.; Watanabe, K.; Taniguchi, T.; Fennie, C.J.; et al. Pressure-controlled interlayer magnetism in atomically thin CrI_3. *Nat. Mater.* **2019**, *18*, 1303–1308. [CrossRef]
68. Song, T.C.; Fei, Z.Y.; Yankowitz, M.; Lin, Z.; Jiang, Q.N.; Hwangbo, K.; Zhang, Q.; Sun, B.S.; Taniguchi, T.; Watanabe, K.; et al. Switching 2D magnetic states via pressure tuning of layer stacking. *Nat. Mater.* **2019**, *18*, 1298–1302. [CrossRef]
69. Wu, Z.W.; Yu, J.; Yuan, S.J. Strain-tunable magnetic and electronic properties of monolayer CrI_3. *Phys. Chem. Chem. Phys.* **2019**, *21*, 7750–7755. [CrossRef]
70. Leon, A.M.; Gonzalez, J.W.; Mejia-Lopez, J.; Lima, F.C.D.; Morell, E.S. Strain-induced phase transition in CrI_3 bilayers. *2D Mater.* **2020**, *7*, 035008. [CrossRef]
71. Vishkayi, S.I.; Torbatian, Z.; Qaiumzadeh, A.; Asgari, R. Strain and electric-field control of spin-spin interactions in monolayer CrI_3. *Phys. Rev. Mater.* **2020**, *4*, 094004. [CrossRef]
72. Safi, A.L.; Chakraborty, S.; Ahmed, M.A.; Chattopadhyay, B. Strain tunable electronic band structure and magnetic anisotropy of CrI_3 bilayer. *ECS J. Solid State Sci. Technol.* **2022**, *11*, 063008. [CrossRef]
73. Li, X.X.; Yang, J.L. $CrXTe_3$ (X = Si, Ge) nanosheets: Two dimensional intrinsic ferromagnetic semiconductors. *J. Phys. Chem. C* **2014**, *2*, 7071–7076. [CrossRef]
74. Zhuang, H.L.; Xie, Y.; Kent, P.R.C.; Ganesh, P. Computational discovery of ferromagnetic semiconducting single-Layer $CrSnTe_3$. *Phys. Rev. B* **2015**, *92*, 035407. [CrossRef]
75. Dong, X.J.; You, J.Y.; Gu, B.; Su, G. Strain-induced room-temperature ferromagnetic semiconductors with large anomalous hall conductivity in two-dimensional $Cr_2Ge_2Se_6$. *Phys. Rev. Appl.* **2019**, *12*, 014020. [CrossRef]
76. Idzuchi, H.; Llacsahuanga Allcca, A.E.; Pan, X.C.; Tanigaki, K.; Chen, Y.P. Increased curie temperature and enhanced perpendicular magneto anisotropy of $Cr_2Ge_2Te_6$/NiO heterostructures. *Appl. Phys. Lett.* **2019**, *115*, 232403. [CrossRef]
77. Zeisner, J.; Alfonsov, A.; Selter, S.; Aswartham, S.; Ghimire, M.P.; Richter, M.; van den Brink, J.; Büchner, B.; Kataev, V. Magnetic anisotropy and spin-polarized two-dimensional electron gas in the van der Waals ferromagnet $Cr_2Ge_2Te_6$. *Phys. Rev. B* **2019**, *99*, 165109. [CrossRef]
78. Khan, I.; Hong, J.S. High curie temperature and strain-induced semiconductor-metal transition with spin reorientation transition in 2D $CrPbTe_3$ monolayer. *Nanotechnology* **2020**, *31*, 195704. [CrossRef]
79. Selter, S.; Bastien, G.; Wolter, A.U.B.; Aswartham, S.; Büchner, B. Magnetic anisotropy and low-field magnetic phase diagram of the quasi-two-dimensional ferromagnet $Cr_2Ge_2Te_6$. *Phys. Rev. B* **2020**, *101*, 014440. [CrossRef]
80. Spachmann, S.; Selter, S.; Büchner, B.; Aswartham, S.; Klingeler, R. Strong uniaxial pressure dependencies evidencing spin-lattice coupling and spin fluctuations in $Cr_2Ge_2Te_6$. *Phys. Rev. B* **2023**, *107*, 184421. [CrossRef]
81. Deiseroth, H.J.; Aleksandrov, K.; Reiner, C.; Kienle, L.; Kremer, R.K. Fe_3GeTe_2 and Ni_3GeTe_2 -two new layered transition-metal compounds: Crystal structures, HRTEM investigations, and magnetic and electrical properties. *Eur. J. Inorg. Chem.* **2006**, *2006*, 1561–1567. [CrossRef]
82. Zhuang, H.L.; Kent, P.R.C.; Hennig, R.G. Strong anisotropy and magnetostriction in the two-dimensional stoner ferromagnet Fe_3GeTe_2. *Phys. Rev. B* **2016**, *93*, 134407. [CrossRef]
83. Joe, M.; Yang, U.; Lee, C.G. First-principles study of ferromagnetic metal Fe_5GeTe_2. *Nano Mater. Sci.* **2009**, *1*, 299–303. [CrossRef]
84. Li, X.L.; Lu, J.T.; Zhang, J.; You, L.; Su, Y.R.; Tsymbal, E.Y. Spin-dependent transport in van der Waals magnetic tunnel junctions with Fe_3GeTe_2 electrodes. *Nano Lett.* **2019**, *19*, 5133–5139. [CrossRef] [PubMed]
85. Hu, X.H.; Zhao, Y.H.; Shen, X.D.; Krasheninnikov, A.V.; Chen, Z.F.; Sun, L.T. Enhanced ferromagnetism and tunable magnetism in Fe_3GeTe_2 monolayer by strain engineering. *ACS Appl. Mater. Inter.* **2020**, *12*, 26367–26373. [CrossRef]
86. Zhu, M.M.; You, Y.R.; Xu, G.Z.; Tang, J.X.; Gong, Y.Y.; Xu, F. Strain modulation of magnetic coupling in the metallic van der Waals magnet Fe_3GeTe_2. *Intermetallics* **2021**, *131*, 107085. [CrossRef]
87. Chen, D.; Sun, W.; Wang, W.X.; Li, X.N.; Li, H.; Cheng, Z.X. Twist-stacked 2D bilayer Fe_3GeTe_2 with tunable magnetism. *J. Mater. Chem. C* **2022**, *10*, 12741–12750. [CrossRef]
88. Ouvrard, G.; Brec, R. Modification of the cationic ordering with respect to the chalcogen in the layered $MM'P_2X_6$ phases. synthesis and structure of two-dimensional $AgVP_2Se_6$. *Mat. Res. Bull.* **1988**, *23*, 1199–1209. [CrossRef]

89. Song, Z.G.; Sun, X.T.; Zheng, J.X.; Pan, F.; Hou, Y.L.; Yung, M.H.; Yang, J.B.; Lu, J. Spontaneous valley splitting and valley pseudospin field effect transistors of monolayer VagP2Se$_6$. *Nanoscale* **2018**, *10*, 13986–13993. [CrossRef]
90. Peng, Y.X.; Cheng, X.; Gu, P.F.; Wang, F.G.; Yang, J.; Xue, M.Z.; Yang, W.Y.; Wang, C.S.; Liu, S.Q.; Watanabe, K.; et al. A quaternary van der Waals ferromagnetic semiconductor AgVP$_2$Se6. *Adv. Funct. Mater.* **2020**, *30*, 1910036. [CrossRef]
91. Joolee, S.; Son, S.; Park, P.; Kim, M.; Tao, Z.; Oh, J.; Lee, T.; Lee, S.; Kim, J.; Zhang, K.X.; et al. Air-stable and layer-dependent ferromagnetism in atomically thin van der Waals $CrPS_4$. *ACS Nano* **2021**, *15*, 16904–16912.
92. Sadhukhan, B.; Bergman, A.; Kvashnin, Y.O.; Hellsvik, J.; Delin, A. Spin-lattice couplings in two-dimensional CrI_3 from first-principles computations. *Phys. Rev. B* **2022**, *105*, 104418. [CrossRef]
93. Kashin, I.V.; Mazurenko, V.V.; Katsnelson, M.I.; Rudenko, A.N. Orbitally-resolved ferromagnetism of monolayer CrI_3. *2D Mater.* **2020**, *7*, 025036. [CrossRef]
94. Soriano, D.; Rudenko, A.N.; Katsnelson, M.I.; Rösner, M. Environmental screening and ligand-field effects to magnetism in CrI_3 monolayer. *NPJ 2D Comput. Mater.* **2021**, *7*, 162. [CrossRef]
95. Esteras, D.L.; Rybakov, A.; Ruiz, A.M.; Baldoví, J.J. Magnon straintronics in the 2D van der Waals ferromagnet CrSBr from first-principles. *Nano Lett.* **2022**, *22*, 8771–8778. [CrossRef]
96. Kvashnin, Y.O.; Bergman, A.; Lichtenstein, A.I.; Katsnelson, M.I. Relativistic exchange interactions in CrX_3 (X=Cl, Br, I) monolayers. *Phys. Rev. B* **2020**, *102*, 115162. [CrossRef]
97. Torelli, D.; Olsen, T. Calculating critical temperatures for ferromagnetic order in two-dimensional materials. *2D Mater.* **2019**, *6*, 015028. [CrossRef]
98. Şaşıoğlu, E.; Sandratskii, L.M.; Bruno, P. First-principles calculation of the intersublattice exchange interactions and curie temperatures of the full heusler alloys Ni_2MnX (X=Ga, In, Sn, Sb). *Phys. Rev. B* **2004**, *70*, 024427. [CrossRef]
99. Singh, S.; Caron, L.; D'Souza, S.W.; Fichtner, T.; Porcari, G.; Fabbrici, S.; Shekhar, C.; Chadov, S.; Solzi, M.; Felser, C. Large magnetization and reversible magnetocaloric effect at the second-order magnetic transition in Heusler materials. *Adv. Mater.* **2016**, *28*, 3321–3325. [CrossRef]
100. Chang, C.; Chen, W.; Chen, Y.; Chen, Y.H.; Chen, Y.; Ding, F.; Fan, C.H.; Fan, H.J.; Fan, Z.; Gong, C.; et al. Recent progress on two-dimensional materials. *Acta Phys.-Chim. Sin.* **2021**, *37*, 2108017. [CrossRef]
101. Ghosh, S.; Ershadrad, S.; Borisov, V.; Sanyal, B. Unraveling effects of electron correlation in two-dimensional Fe_nGeTe_2 (N = 3, 4, 5) by dynamical mean field theory. *NPJ 2D Comput. Mater.* **2023**, *9*, 86. [CrossRef]
102. Kim, K.S.; Zhao, Y.; Jang, H.; Lee, S.Y.; Kim, J.M.; Kim, K.S.; Ahn, J.H.; Kim, P.; Choi, J.Y.; Hong, B.H. Large-scale pattern growth of graphene films for stretchable transparent electrodes. *Nature* **2009**, *457*, 706–710. [CrossRef]
103. Liu, J.Y.; Sun, Q.; Kawazoe, Y.; Jena, P. Exfoliating biocompatible ferromagnetic Cr-trihalide monolayers. *Phys. Chem. Chem. Phys.* **2016**, *13*, 8777–8784. [CrossRef] [PubMed]
104. Guo, H.Y.; Lu, N.; Wang, L.; Wu, X.J.; Zeng, X.C. Tuning electronic and magnetic properties of early transition-metal dichalcogenides via tensile strain. *J. Phys. Chem. C* **2014**, *118*, 7242–7249. [CrossRef]
105. Lv, H.Y.; Lu, W.J.; Shao, D.F.; Liu, Y.; Sun, Y.P. Strain-controlled switch between ferromagnetism and antiferromagnetism in CrSeTe monolayers. *Phys. Rev. B* **2015**, *92*, 214419. [CrossRef]
106. Liu, Y.H.; Kwon, S.; de Coster, G.J.; Lake, R.K.; Neupane, M.R. Structural, electronic, and magnetic properties of $CrTe_2$. *Phys. Rev. Mater.* **2022**, *6*, 084004. [CrossRef]
107. Fumega, A.O.; Phillips, J.; Victor Pardo, V. Controlled two-dimensional ferromagnetism in 1T–$CrTe_2$: The role of charge density wave and strain. *J. Phys. Chem. C* **2020**, *124*, 21047–21053. [CrossRef]
108. Song, Y.; Wang, X.C.; Mi, W.B. Role of electron filling in the magnetic anisotropy of monolayer WSe_2 doped with 5*d* transition metals. *Phys. Rev. Mater.* **2017**, *1*, 074408. [CrossRef]
109. Liu, J.; You, Y.R.; Batashev, I.; Gong, Y.Y.; You, X.M.; Huang, B.W.; Zhang, F.Q.; Miao, X.F.; Xu, F.; van Dijk, N.; et al. Design of reversible low-field magnetocaloric effect at room temperature in hexagonal Mn Ferromagnets. *Phys. Rev. Appl.* **2020**, *13*, 054003. [CrossRef]
110. Stahl, J.; Shlaen, E.; Johrendt, D. The van der Waals ferromagnets $Fe_{5-\delta}GeTe_2$ and $Fe_{5-\delta-X}Ni_xGeTe_2$ -crystal structure, stacking faults, and magnetic properties. *Z. Anorg. Allg. Chem.* **2018**, *644*, 1923–1929. [CrossRef]
111. May, A.F.; Ovchinnikov, D.; Zheng, Q.; Hermann, R.; Calder, S.; Huang, B.V.; Fei, Z.Y.; Liu, Y.H.; Xu, X.D.; McGuire, M.A. Ferromagnetism near room temperature in the cleavable van der Waals crystal Fe_5GeTe_2. *ACS Nano* **2019**, *13*, 4436–4442. [CrossRef] [PubMed]
112. Dhakal, K.P.; Roy, S.; Jang, H.; Chen, X.; Yun, W.S.; Kim, H.; Lee, J.D.; Kim, J.; Ahn, J.H. Local strain induced band gap modulation and photoluminescence enhancement of multilayer transition metal dichalcogenides. *Chem. Mater.* **2017**, *29*, 5124–5133. [CrossRef]
113. Iguiniz, N.; Frisenda, R.; Bratschitsch, R.; Castellanos-Gomez, A. Revisiting the buckling metrology method to determine the Young's modulus of 2D materials. *Adv. Mater.* **2019**, *31*, 1807150. [CrossRef]
114. Liu, Z.; Amani, M.; Najmaei, S.; Xu, Q.; Zou, X.L.; Zhou, W.; Yu, T.; Qiu, C.Q.; Glen Birdwell, A.; Crowne, F.J.; et al. Strain and structure heterogeneity in MoS_2 atomic layers grown by chemical vapour deposition. *Nat. Commun.* **2014**, *5*, 5246. [CrossRef]
115. Brennan, C.J.; Nguyen, J.; Yu, E.T.; Lu, N.S. Interface adhesion between 2D materials and elastomers measured by buckle delaminations. *Adv. Mater. Interfaces* **2015**, *2*, 1500176. [CrossRef]

116. Plechinger, G.; Castellanos-Gomez, A.; Buscema, M.; van der Zant, H.S.J.; Steele, G.A.; Kuc, A.; Heine, T.; Schüller, C.; Korn, T. Control of biaxial strain in single-layer molybdenite using local thermal expansion of the substrate. *2D Mater.* **2015**, *2*, 015006. [CrossRef]
117. Yang, R.; Lee, J.; Tang, G.H.; Sankaran, R.M.; Zorman, C.A.; Feng, P.X.L. Tuning optical signatures of single- and few-layer MoS_2 by blown-bubble bulge straining up to fracture. *Nano Lett.* **2017**, *17*, 4568–4575. [CrossRef]
118. Pak, S.; Lee, J.; Lee, Y.W.; Jang, A.R.; Ahn, S.; Ma, K.Y.; Cho, Y.; Hong, J.; Lee, S.; Jeong, H.Y.; et al. Strain-mediated interlayer coupling effects on the excitonic behaviors in an epitaxially grown MoS_2/WS_2 van der Waals heterobilayer. *Nano Lett.* **2017**, *17*, 5634–5640. [CrossRef]
119. John, A.P.; Thenapparambil, A.; Thalakulam, M. Strain-engineering the schottky barrier and electrical transport on MoS_2. *Nanotechnology* **2020**, *31*, 275703. [CrossRef] [PubMed]
120. Thai, K.Y.; Park, I.J.; Kim, B.J.; Hoang, A.T.; Na, Y.; Park, C.U.; Chae, Y.; Ahn, J.H. MoS_2/graphene photodetector array with strain-modulated photoresponse up to the near-infrared regime. *ACS Nano* **2021**, *15*, 12836–12846. [CrossRef] [PubMed]
121. Hu, L.; Zhou, J.; Hou, Z.P.; Su, W.T.; Yang, B.Z.; Li, L.W.; Yan, M. Polymer-buried van der Waals magnets for promising wearable room-temperature spintronics. *Mater. Horiz.* **2021**, *8*, 3306–3314. [CrossRef] [PubMed]
122. Li, Z.W.; Lv, Y.W.; Ren, L.W.; Li, J.; Kong, L.G.; Zeng, Y.J.; Tao, Q.Y.; Wu, R.X.; Ma, H.F.; Zhao, B.; et al. Efficient strain modulation of 2D materials via polymer encapsulation. *Nat. Commun.* **2020**, *11*, 1151. [CrossRef] [PubMed]
123. Tian, C.K.; Pan, F.H.; Xu, S.; Ai, K.; Xia, T.L.; Cheng, P. Tunable magnetic properties in van der Waals crystals $(Fe_{1-X}Co_X)_5GeTe_2$. *Appl. Phys. Lett.* **2019**, *115*, 232403. [CrossRef]
124. Seo, J.; Kim, D.Y.; An, E.S.; Kim, K.; Kim, G.Y.; Hwang, S.Y.; Kim, D.W.; Jang, B.G.; Kim, H.; Eom, G.; et al. room temperature ferromagnetism in a magnetic metal-rich van der Waals metal. *Sci. Adv.* **2020**, *6*, eaay8912. [CrossRef] [PubMed]
125. Zhang, L.M.; Song, L.Y.; Dai, H.W.; Yuan, J.H.; Wang, M.S.; Huang, X.Y.; Qiao, L.; Cheng, H.; Wang, X.; Ren, W.; et al. Substrate-modulated ferromagnetism of two-dimensional Fe_3GeTe_2. *Appl. Phys. Lett.* **2020**, *116*, 042402. [CrossRef]
126. Deng, Y.J.; Yu, Y.J.; Song, Y.C.; Zhang, J.Z.; Wang, N.Z.; Sun, Z.Y.; Yi, Y.F.; Wu, Y.Z.; Wu, S.W.; Zhu, J.Y.; et al. Gate-tunable room-temperature ferromagnetism in two-dimensional Fe_3GeTe_2. *Nature* **2018**, *563*, 94–99. [CrossRef]
127. Li, Q.; Yang, M.M.; Gong, C.; Chopdekar, R.V.; N'Diaye, A.T.; Turner, J.; Chen, G.; Scholl, A.; Shafer, P.; Arenholz, E.; et al. Patterning-induced ferromagnetism of Fe_3GeTe_2 van der Waals materials beyond room temperature. *Nano Lett.* **2018**, *18*, 5974–5980. [CrossRef]
128. Yang, M.M.; Li, Q.; Chopdekar, R.V.; Stan, C.; Cabrini, S.; Choi, J.W.; Wang, S.; Wang, T.Y.; Gao, N.; Scholl, A.; et al. Highly enhanced curie temperature in Ga-implanted Fe_3GeTe_2 van der Waals material. *Adv. Quantum Technol.* **2020**, *3*, 2000017. [CrossRef]
129. Niu, W.; Zhang, X.Q.; Wang, W.; Sun, J.B.; Xu, Y.B.; He, L.; Liu, W.Q.; Pu, Y. Probing the atomic-scale ferromagnetism in van der Waals magnet $CrSiTe_3$. *Appl. Phys. Lett.* **2021**, *119*, 172402. [CrossRef]
130. Kim, D.; Park, S.; Lee, J.; Yoon, J.; Joo, S.; Kim, T.; Min, K.J.; Park, S.Y.; Kim, C.; Moon, K.W.; et al. Antiferromagnetic coupling of van der Waals ferromagnetic Fe_3GeTe_2. *Nanotechnology* **2019**, *30*, 245701. [CrossRef]
131. Ribeiro, M.; Gentile, G.; Marty, A.; Dosenovic, D.; Okuno, H.; Vergnaud, C.; Jacquot, J.F.; Jalabert, D.; Longo, D.; Ohresser, P.; et al. Large-scale epitaxy of two-dimensional van der Waals room-temperature ferromagnet Fe_5GeTe_2. *NPJ 2D Mater. Appl.* **2022**, *6*, 10. [CrossRef]
132. Dong, X.J.; You, J.Y.; Zhang, Z.; Gu, B.; Su, G. Great enhancement of curie temperature and magnetic anisotropy in two-dimensional van der Waals magnetic semiconductor heterostructures. *Phys. Rev. B* **2020**, *102*, 144443. [CrossRef]
133. Carteaux, V.; Brunet, D.; Ouvrard, G.; Andre, G. Crystallographic, magnetic and electronic structures of a new layered ferromagnetic compound $Cr_2Ge_2Te_6$. *J. Phys. Condens. Matter* **1995**, *7*, 69. [CrossRef]
134. Spachmann, S.; Elghandour, A.; Selter, S.; Büchner, B.; Aswartham, S.; Klingeler, R. Strong effects of uniaxial pressure and short-range correlations in $Cr_2Ge_2Te_6$. *Phys. Rev. Res.* **2022**, *4*, L022040. [CrossRef]
135. Wang, J.W.; Han, M.J.; Wang, Q.; Ji, Y.Q.; Zhang, X.; Shi, R.; Wu, Z.F.; Zhang, L.; Amini, A.; Guo, L.; et al. Strained epitaxy of monolayer transition metal dichalcogenides for wrinkle arrays. *ACS Nano* **2021**, *15*, 6633–6644. [CrossRef] [PubMed]
136. Wang, E.Z.; Chen, Z.K.; Shi, R.; Xiong, Z.X.; Xin, Z.X.; Wang, B.L.; Guo, J.; Peng, R.X.; Wu, Y.H.; Li, C.Y.; et al. Humidity-controlled dynamic engineering of buckling dimensionality in MoS_2 thin films. *ACS Nano* **2022**, *16*, 14157–14167. [CrossRef]
137. Wang, J.W.; Luo, Y.; Cai, X.B.; Shi, R.; Wang, W.J.; Li, T.R.; Wu, Z.F.; Zhang, X.; Peng, O.W.; Amini, A.; et al. Multiple regulation over growth direction, band structure, and dimension of monolayer WS_2 by a quartz substrate. *Chem. Mater.* **2020**, *32*, 2508–2517. [CrossRef]
138. Wang, E.Z.; Xiong, Z.X.; Chen, Z.K.; Xin, Z.Q.; Ma, H.C.; Ren, H.T.; Wang, B.L.; Guo, J.; Sun, Y.F.; Wang, X.W.; et al. Water nanolayer facilitated solitary-wave-like blisters in MoS_2 thin films. *Nat. Commun.* **2023**, *14*, 4324. [CrossRef]
139. Ren, H.T.; Xiong, Z.X.; Wang, E.Z.; Yuan, Z.Q.; Sun, Y.F.; Zhu, K.L.; Wang, B.L.; Wang, X.W.; Ding, H.Y.; Liu, P.; et al. Watching dynamic self-assembly of web buckles in strained MoS_2 thin films. *ACS Nano* **2019**, *13*, 3106–3116. [CrossRef]
140. Ahn, G.H.; Amani, M.; Rasool, H.; Lien, D.H.; Mastandrea, J.P.; Ager, J.W.; Dubey, M.; Chrzan, D.C.; Minor, A.M.; Javey, A. Strain-engineered growth of two-dimensional materials. *Nat. Commun.* **2017**, *8*, 608. [CrossRef]
141. Li, H.; Qi, X.; Wu, J.; Zeng, Z.; Wei, J.; Zhang, H. Investigation of MoS_2 and graphene nanosheets by magnetic force microscopy. *ACS Nano* **2013**, *7*, 2842–2849. [CrossRef] [PubMed]

142. Kezilebieke, S.; Silveira, O.J.; Huda, M.N.; Vaňo, V.; Aapro, M.; Ganguli, S.C.; Lahtinen, J.; Mansell, R.; Dijken, S.V.; Foster, A.S.; et al. Electronic and magnetic characterization of epitaxial $CrBr_3$ monolayers on a superconducting substrate. *Adv. Mater.* **2021**, *33*, 2006850. [CrossRef] [PubMed]
143. Kerr, J.X. On rotation of the plane of polarization by reflection from the pole of a magnet. *London Edinb. Dublin Philos. Mag. J. Sci.* **1877**, *3*, 321–343. [CrossRef]
144. Wang, Z.; Zhang, T.Y.; Ding, M.; Dong, B.J.; Li, Y.X.; Chen, M.L.; Li, X.X.; Huang, J.Q.; Wang, H.W.; Zhao, X.T.; et al. Electric-field control of magnetism in a few-layered van der Waals ferromagnetic semiconductor. *Nat. Nanotechnol.* **2018**, *13*, 554–559. [CrossRef]
145. Fei, Z.Y.; Huang, B.; Malinowski, P.; Wang, W.B.; Song, T.C.; Sanchez, J.; Yao, W.; Xiao, D.; Zhu, X.Y.; May, A.F.; et al. Two-dimensional itinerant ferromagnetism in atomically thin Fe_3GeTe_2. *Nat. Mater.* **2018**, *17*, 778–782. [CrossRef] [PubMed]
146. Jin, C.H.; Tao, Z.; Kang, K.F.; Watanabe, K.; Taniguchi, T.; Mak, K.F.; Shan, J. Imaging and control of critical fluctuations in two-dimensional magnets. *Nat. Mater.* **2020**, *19*, 1290–1294. [CrossRef]
147. Jiang, S.W.; Shan, J.; Mak, K.F. Electric-field switching of two-dimensional van der Waals magnets. *Nat. Mater.* **2018**, *17*, 406–410. [CrossRef] [PubMed]
148. Huang, B.; Clark, G.; Klein, D.R.; MacNeill, D.; Navarro-Moratalla, E.; Seyler, K.L.; Wilson, N.; McGuire, M.A.; Cobden, D.H.; Xiao, D.; et al. Electrical control of 2D magnetism in bilayer CrI_3. *Nat. Nanotechnol.* **2018**, *13*, 544–548. [CrossRef]
149. Jiang, S.W.; Li, L.Z.; Wang, Z.F.; Mak, K.F.; Jie Shan, J. Controlling magnetism in 2D CrI_3 by electrostatic doping. *Nat. Nanotechnol.* **2018**, *13*, 549–553. [CrossRef]
150. Ding, B.; Li, Z.F.; Xu, G.Z.; Li, H.; Hou, Z.P.; Liu, E.K.; Xi, X.K.; Xu, F.; Yao, Y.; Wang, W.H. Observation of magnetic skyrmion bubbles in a van der Waals ferromagnet Fe_3GeTe_2. *Nano Lett.* **2020**, *20*, 868–873. [CrossRef]
151. Park, T.E.; Peng, L.C.; Liang, J.H.; Hallal, A.; Yasin, F.S.; Zhang, X.C.; Song, K.M.; Kim, S.J.; Kim, K.; Weigand, M.; et al. Neel-type skyrmions and their current-induced motion in van der Waals ferromagnet-based heterostructures. *Phys. Rev. B* **2021**, *104410*, 024427. [CrossRef]
152. Wang, H.; Wang, C.X.; Li, Z.A.; Tian, H.F.; Shi, Y.G.; Yang, H.X.; Li, J.Q. Characteristics and temperature-field-thickness evolutions of magnetic domain structures in van der Waals magnet Fe_3GeTe_2 nanolayers. *Appl. Phys. Lett.* **2020**, *116*, 192403. [CrossRef]
153. Chen, W.J.; Sun, Z.Y.; Wang, Z.J.; Gu, L.H.; Xu, X.D.; Wu, S.W.; Gao, C.L. Direct observation of van der Waals stacking–dependent interlayer magnetism. *Science* **2019**, *366*, 983–987. [CrossRef] [PubMed]
154. Schmid, I.; Marioni, M.A.; Kappenberger, P.; Romer, S.; Parlinska-Wojtan, M.; Hug, H.J.; Hellwig, O.; Carey, M.J.; Fullerton, E.E. Exchange bias and domain evolution at 10 nm scales. *Phys. Rev. Lett.* **2010**, *105*, 197201. [CrossRef]
155. Kirtley, J.R.; Paulius, L.; Rosenberg, A.J.; Palmstrom, J.C.; Holland, C.M.; Spanton, E.M.; Schiessl, D.; Jermain, C.L.; Gibbons, J.; Fung, Y.K.K.; et al. Scanning SQUID susceptometers with sub-micron spatial resolution. *Rev. Sci. Instrum.* **2016**, *87*, 093702. [CrossRef] [PubMed]
156. Finkler, A.; Vasyukov, D.; Segev, Y.; Ne'eman, L.; Lachman, E.O.; Rappaport, M.L.; Myasoedov, Y.; Zeldov, E.; Huber, M.E. Scanning superconducting quantum interference device on a tip for magnetic imaging of nanoscale phenomena. *Rev. Sci. Instrum.* **2012**, *83*, 073702. [CrossRef]
157. Vasyukov, D.; Anahory, Y.; Embon, L.; Halbertal, D.; Cuppens, J.; Neeman, L.; Finkler, A.; Segev, Y.; Myasoedov, Y.; Rappaport, L.; et al. A scanning superconducting quantum interference device with single electron spin sensitivity. *Nat. Nanotechnol.* **2013**, *8*, 639–644. [CrossRef]
158. Bagani, K.; Sarkar, J.; Uri, A.; Rappaport, M.L.; Huber, M.E.; Zeldov, E.; Myasoedov, Y. Sputtered $Mo_{66}Re_{34}$ squid-on-tip for high-field magnetic and thermal nanoimaging. *Phys. Rev. Appl.* **2019**, *12*, 044062. [CrossRef]
159. Ren, H.T.; Zhong, J.; Xiang, G. The progress on magnetic material thin films prepared using polymer-assisted deposition. *Molecules* **2023**, *28*, 5004. [CrossRef]
160. Chang, K.; Eichler, A.; Rhensius, J.; Lorenzelli, L.; Degen, C.L. Nanoscale imaging of current density with a single-spin magnetometer. *Nano Lett.* **2017**, *17*, 2367–2373. [CrossRef]
161. Balasubramanian, G.; Chan, I.Y.; Kolesov, R.; Al-Hmoud, M.; Tisler, J.; Shin, C.; Kim, C.; Wojcik, A.; Hemmer, P.R.; Krueger, A.; et al. Nanoscale imaging magnetometry with diamond spins under ambient conditions. *Nature* **2008**, *455*, 648–651. [CrossRef] [PubMed]
162. Ariyaratne, A.; Bluvstein, D.; Myers, B.A.; Bleszynski Jayich, A.C. Nanoscale electrical conductivity imaging using a nitrogen-vacancy center in diamond. *Nat. Commun.* **2019**, *9*, 2406. [CrossRef] [PubMed]
163. Vool, U.; Hamo, A.; Varnavides, G.; Wang, Y.; Zhou, T.X.; Kumar, N.; Dovzhenko, Y.; Qiu, Z.; Garcia, C.A.C.; Pierce, A.T. Imaging phonon-mediated hydrodynamic flow in WTe_2. *Nat. Phys.* **2021**, *17*, 1216–1220. [CrossRef]
164. Wörnle, M.S.; Welter, P.; Giraldo, M.; Lottermoser, T.; Fiebig, M.; Gambardella, P.; Degen, C.L. Coexistence of Bloch and N'eel walls in a collinear antiferromagnet. *Phys. Rev. B* **2021**, *103*, 094426. [CrossRef]

nanomaterials

MDPI

Review

Recent Advances in the Spintronic Application of Carbon-Based Nanomaterials

Shweta Pawar [1,2], Hamootal Duadi [1] and Dror Fixler [1,2,*]

[1] Faculty of Engineering and the Institute of Nanotechnology and Advanced Materials, Bar Ilan University, Ramat Gan 5290002, Israel

[2] Bar-Ilan Institute of Nanotechnology & Advanced Materials (BINA), Bar Ilan University, Ramat Gan 5290002, Israel

* Correspondence: dror.fixler@biu.ac.il; Tel.: +97-23-531-7598

Abstract: The term "carbon-based spintronics" mostly refers to the spin applications in carbon materials such as graphene, fullerene, carbon nitride, and carbon nanotubes. Carbon-based spintronics and their devices have undergone extraordinary development recently. The causes of spin relaxation and the characteristics of spin transport in carbon materials, namely for graphene and carbon nanotubes, have been the subject of several theoretical and experimental studies. This article gives a summary of the present state of research and technological advancements for spintronic applications in carbon-based materials. We discuss the benefits and challenges of several spin-enabled, carbon-based applications. The advantages include the fact that they are significantly less volatile than charge-based electronics. The challenge is in being able to scale up to mass production.

Keywords: carbon-based spintronics; graphene; carbon nanotube; fullerenes; carbon nitride

check for updates

Citation: Pawar, S.; Duadi, H.; Fixler, D. Recent Advances in the Spintronic Application of Carbon-Based Nanomaterials. *Nanomaterials* **2023**, *13*, 598. https://doi.org/10.3390/nano13030598

Academic Editors: Oleksandr Dobrovolskiy, Gang Xiang and Hongtao Ren

Received: 9 January 2023
Revised: 29 January 2023
Accepted: 30 January 2023
Published: 2 February 2023

1. Introduction

In the 1990s, the term spintronics (short for spin electronics) was created to designate technologies that make use of an electron's quantum mechanical feature known as "spin", which has only two values: spin-up and spin-down. To keep up with the ever-increasing data generation, our society has dramatically boosted data storage capacity. Memory devices have shrunk in size concurrently [1]. In part, this growth in data storage capacity at smaller dimensions makes it easier to carry and exchange online material such as music and photographs on mobile devices. This field of study and technology is frequently referred to as spintronics since a significant portion of this technological breakthrough has been made possible by controlling the spin degree of freedom of electrons in circuits. After the giant magnetoresistance (GMR) effect was discovered in the 1980s, spintronics research developed pace [2]. One of the newer technologies that will allow next-generation nanoelectronics devices to have more memory and processing capability while using less electricity is spintronics. Such devices make use of the degree of freedom that electrons and/or holes' spins have, which can also interact with their orbital moments. The spin polarization in these devices is managed either by magnetic layers acting as spin-polarizers or spin-polarizer/analyzers or by spin–orbit coupling [3].

The carbon material research focuses on carbon nanotubes and graphene. Although carbon exists in a variety of allotropic forms, these two substances offer a low dimensionality and high crystalline purity, opening the door to several unusual quantum events [4]. Spin injection and transport in carbon materials, such as carbon nanotubes, graphene, fullerene, and organic materials, are referred to as "carbon-based spintronics". They can display strong long-range magnetic ordering and have much applicability if magnetic moments are introduced. For application in computers, tablets, and handheld devices, spintronics is utilized in fast microchips, logic gates, transistors, and capacitors. Additionally, all-hydrocarbon products can drastically cut the demand for crucial materials used in the semiconductor sector (e.g., indium and gallium) [3,5]. It is necessary to comprehend the history of magnetic coupling and make improvements to this vital characteristic to create new and significantly improved spintronic devices. At the same time, traditional research into hybrids of pure carbon units and magnetic metal species, known as magnetically functionalized carbon nanostructures, continues to advance rapidly. When creating composites with tunable magnetic characteristics, carbon components are essential because they serve as templates for magnetic units. Various kinds of carbon nanomaterials are utilized to shield and stabilize encapsulated magnetic subunits by forming cages with high cohesion [6–8].

To properly understand carbon-based spintronics, it is crucial to assess the most recent research in the area. This review article has explored the evolution of carbon-based spintronics and has also discussed some current problems and anticipated future developments. The preparation of the materials needed for carbon-based spintronic is covered in the first section. The second section illustrates the operation of carbon-based spintronic device operations. In this section, a few potential carbon-based spintronic devices for logic and memory applications are introduced along with other uses. The last section discusses the topic of prospective and challenges in carbon-based spintronics for real-world applications.

2. Carbon-Based Spintronics

2.1. General

The different carbon prototypes used in the synthesis serve as the foundation for most carbon-based spintronic devices. The carbon materials utilized are divided into four distinct types: graphene-based, carbon nanotubes-based, carbon nitride-based, and fullerenes-based (Figure 1). These are representations of carbon's nanoscopic allotropes in two, one, and zero dimensions.

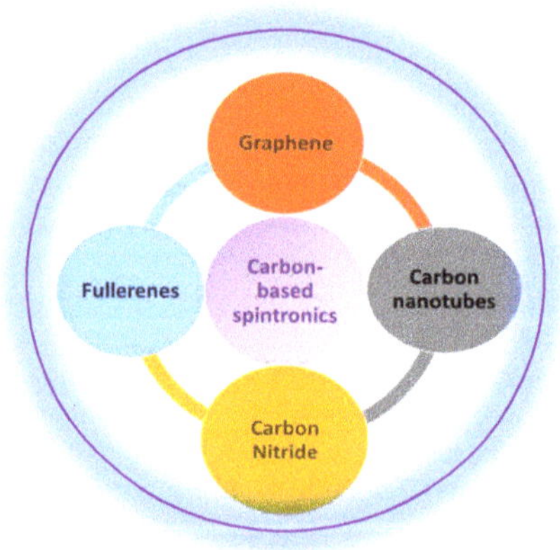

Figure 1. Schematic representation of carbon-based spintronics.

2.2. Preparation

It is possible to think about graphene as a single sheet of graphite since it produces a perfectly planar layer of carbon atom. Thus, the arrangement of carbon atoms results in a hexagonally symmetric, monatomic, two-dimensional lattice. Currently, there are three main methods for creating graphene nanoribbons (GNRs): cutting graphene using lithography, polycyclic bottom-up synthesis, and unzipping carbon nanotubes (CNTs). Chemical vapor deposition (CVD) on metal surfaces, mechanical exfoliation, electrochemical exfoliation, chemical exfoliation of graphite powder, epitaxial growth on single crystal SiC, chemical coupling processes, and intercalation/sonication are the common techniques used to produce graphene nanosheets (GNs) [9]. Graphene quantum dots (GQDs) can be created using top-down or bottom-up methods. In the top-down strategy, different synthetic and fabrication techniques are used to create zero-dimensional GQDs from specific carbon source materials (carbon fibers, graphene oxide, and graphene nanoribbon) [10]. These include microwave-assisted hydrothermal, solvothermal, chemical oxidation, microfluidization, and electrochemical techniques. Using a bottom-up strategy, stepwise solution chemistry, pyrolysis and oxidation, and hydrothermal heating are used to combine smaller benzene derivative units into larger GQD entities. The single-step Wurtz reaction is used to create scalable carbon nanosheets (CNs) without magnetic impurities, which feature an amorphous structure with embedded crystalline graphene nanocrystals by Liu C. et al. [11]. The synthesis is depicted in Figure 2.

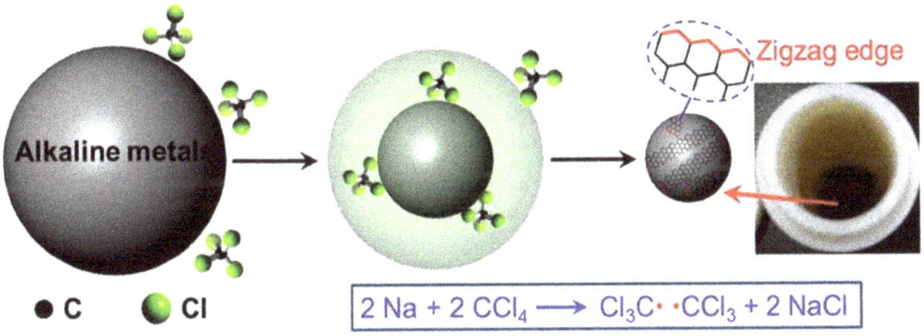

Figure 2. Synthesis of ferromagnetic carbon nanosheets. Reproduced from ref [11] with permission of Journal of Physical Chemistry C copyright 2020.

Arc discharge [12], chemical vapor deposition, and laser ablations are some of the techniques used to produce CNTs [13]. Most frequently, reactive nitrogen-rich, oxygen-free molecules with pre-bonded C-N core structures, such as triazine and heptazine derivatives, are utilized as precursors in the chemical production of gC3N4. Using conveniently accessible nitrogen-rich precursors such as cyanamide, urea, thiourea, melamine, and dicyandiamide, it can be produced with just one step of heat processing [14]. Initially, fullerenes were created by laser vaporizing carbon in an inert atmosphere, but this process only yielded trace amounts of fullerenes. Later on, using the arc heating of graphite and the laser irradiation of polyaromatic hydrocarbon (PAHs), substantial amounts of fullerene C60 were produced. The synthesis process for different carbon-based materials is illustrated in Figure 3.

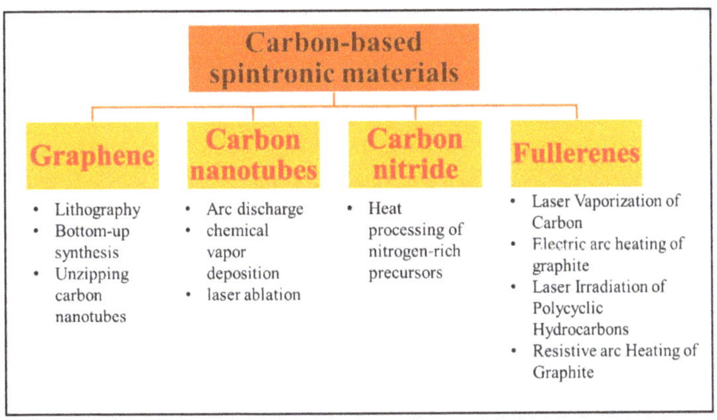

Figure 3. Schematic representation of the preparation of carbon-based spintronic materials.

3. Applications

3.1. Graphene-Based

This carbon-based substance, graphene, is fundamentally important in spintronics and the reason for this decision was due to its inherent magnetic properties. The dimensionally reduced graphene nanoribbons are introduced in this section along with a general description of graphene. The zigzag varieties of these species, which have been shown to display magnetic properties in their ground state, can be viewed as the basic elements of carbon nanostructure magnetism. With a 1.97 eV optical bandgap, carbon nano solenoid (CNS) exhibits strong red photoluminescence. According to the results of magnetic tests, the CNS exhibits complicated magnetic ordering behavior and a paramagnetism response at low temperatures. Overall, such a Π-conjugated CNS makes it possible to serve as the foundation for the creation of electrical and spintronic devices that comprise of CNS molecules [15]. Partially hydrogenated graphene (PHGr), produced by S. Yang and his research team, contains periodic hexagonal graphene nanoflakes (GNFs) with zigzag borders embedded within a layer of hydrogenated graphene [16]. They revealed that a single Au atom can catalyze H_2 dissociation by selectively adhering to borders, lowering the barrier to graphene hydrogenation. The PHGr can therefore be used as a platform for single-atom catalysts and carbon-based spintronic devices. Additionally, when the size of the GNF fluctuates between 1.4 nm and 2.3 nm, it was discovered that the antiferromagnetic boundary states and semiconducting characteristics remain unaltered. The impact of magnetic coupling engineering on the spin transport characteristics of covalently coupled nanographene dimer-based molecular devices has been investigated by Jing Zeng and Xiaohui Deng. Their findings demonstrate that the magnetic coupling strength of the covalently connected, five-membered ring-decorated nanographene is a key factor in the superior spin transport capabilities [17]. The on-surface synthesis of covalently bound triangulene dimers and a proof-of-concept experimental analysis of magnetism are presented by Mishra S. et al. The selective synthesis of triangulene dimers, in which the triangulene units are either directly joined through their minority sublattice atoms or are separated by a 1,4-phenylene spacer, results from on-surface interactions of rationally chosen precursor molecules on Au(111) [18]. Bond-resolved scanning tunneling microscopy has been used to describe the chemical composition of the dimers. Collective singlet–triplet spin excitations are discovered in the dimers by scanning tunneling spectroscopy and inelastic electron tunneling spectroscopy experiments, proving effective intertriangulene magnetic coupling. Using 2,6-dibromo-1,5-diphenylnaphthalene as a precursor, Keerthi A. et al. describe the on-surface synthesis of an unheard-of cove-edge chiral GNR with a benzo-fused backbone on an Au(111) surface [19]. A scanning tunneling microscopy and spectroscopy indicate the initial precursor self-assembly and the development of the chiral GNRs following annealing, as well as a very tiny electronic bandgap of about 1.6 eV.

The "inverse Schottky" feature, high spin polarization, the reverse of spin polarization, and substantial magnetoresistance were among the remarkable spin transport properties discovered. These findings suggest that the covalently bonded, five-membered ring-decorated nanographene dimers can be engineered to function as overcurrent protection, information storage, and logic devices. The research supports a practical approach to producing high-performance molecular spintronic devices made of carbon. Different emergent spin-related quantum phenomena are strongly influenced by spin–orbit coupling (SOC). Additionally, light carbon atoms imply a small intrinsic SOC strength, which prevents their use in spintronics. Gu J. et al. develop a specific deformation vector with chiral curvature, to replicate the warping and twisting of space to increase the SOC effect [20]. This opens the door to the creation of numerous spintronic devices, such as spin current sources, spin field-effect transistors, spin galvanic meters, and inverse spin Hall effect-related spin current sources and transistors. The discovery by de Sousa, M.S.M. et al. that periodic vacancies can transform single-layer graphene into a nodal-line or nodal-loop semimetal opens the door to the creation of novel capabilities for spintronic or electrical devices based on graphene [21]. Additionally, the density of states at the Fermi level is increased in the vacancy-engineered graphene. This method is true even in the presence of high spin–orbit coupling, regardless of the specifics of the systems, and applies to a variety of crystalline structures. The isolated cyclo [18] carbon (C18), a recently synthesized carbon allotrope, has the potential to be the essential element of advanced electronic systems because of its superior physical characteristics. Based on the non-equilibrium Green's function approach and density functional theory, Hou L. et al. studied a unique molecular spin-filtering device that was created by covalently sandwiching a C18 ring between two zigzag-edged graphene nanoribbons (ZGNRs) [22]. By sandwiching a metal-carbon chain between C18 and nonmagnetic ZGNRs (on/off ratio of $\approx$2737), it is possible to produce a potent bi-molecular configuration switching effect. Correa J.D. et al. systematically investigate the electrical and magnetic properties of a class of Penta-graphene-related materials that are produced by swapping out the four-fold coordinated carbon atoms for other elements [23]. Pentagonal nanoribbons exhibit remarkable electrical and magnetic properties as a result of quantum confinement and edge effects. These systems can retain magnetic states with different electronic behaviors or spin-unpolarized states, depending on the particular pentagonal material and edge geometries. Liang Z.'s group uses density functional theory and non-equilibrium Green's function to conduct a theoretical study on the electrical characteristics of double atomic carbon chains bridging graphene electrodes. Atomic carbon chains' conductivity is significantly impacted by strain. But the intrinsic transport of double atomic carbon chains is mostly governed by the coupling effect between adjacent chains. The connected double atomic chains have strong antiparallel spin-filtering capabilities on two electrodes. In spintronic devices and carbon-based field-effect transistors, the connected double-atomic carbon chains have a wide range of possible applications. An all-carbon spintronic device made of a perylene molecule connected to two symmetrical ferromagnetic zigzag-edge graphene nanoribbon (ZGNR) electrodes via carbon atomic chains was the subject of an investigation by Han X. et al. [24]. The ZGNR electrodes magnetization could be changed by applying an external magnetic field. The setup displays spin filtering and the negative differential resistance (NDR) effect when the spin configuration is parallel. Antiparallel spin arrangement reveals bipolar spin filtering, spin rectifying, and a NDR effect. Additionally, there is a significant magnetoresistance difference between the parallel and antiparallel spin arrangements. By organizing the direction of the magnetic moments of carbon atoms at the edges, Prayitno T.B., Budi E. and Fahdiran R. have shown how first-principles calculations may be used to modify the band gap of the bilayer zigzag graphene nanoribbon [25]. The polar angle, as it is specified in spherical coordinates, was used to specify these directions. The band gap increases as the polar angle rises from the ferromagnetic configuration to the antiferromagnetic state. In addition, it has been demonstrated that the ferromagnetic configuration leads to the metallic system, whereas the others lead to the insulator. A straightforward plasma technique was used by Jeong et al. to create N-doped graphene, and the resulting material was shown to have a specific capacitance of 280 F/g, which is four times greater than that of the similar undoped pure graphene [26]. This is due to the fact that N-doping can improve the electrical conductivity of graphene and introduce charge-transferring sites through a doping-induced charge modulation, leading to an increase in the specific capacitance and an improvement in power density of 8105 W/kg and energy density of 48 Wh/kg. GO can also be hydrothermally reduced using compounds that contain nitrogen to create N-doped graphene.

3.1.1. Doping or Embedding

The structure of nanoscale carbon materials can be improved and controlled through doping, which in turn affects their optical, electrical, dielectric, and magnetic properties. It was shown that the system's spin characteristics can be dramatically affected by cobalt (Co) atoms chemisorbed on p-conjugated C atoms. Distinct Co atoms have different distributions of the spin densities of the various energy levels, creating pathways for effective spin-transfer operations. Both global spin-transfer activities between the Co atoms and reversible local spin-flip events on each Co atom are accomplished [27]. Zhang P.

and colleagues showed that the magnetism in boron atoms (B2) doped seven-atom-wide armchair graphene nanoribbons (B2-7AGNRs) is provided by p-electrons, resulting from the imbalance of electrons in two spin channels because of boron dopants [28]. The significant observation of R. Langer and his team was that doping graphene with two separate transition metal atoms and creating transition metal dimers results in a vastly increased magnetic anisotropy energy (MAE) when compared to graphene doped with a single atom [29]. A new two-dimensional carbon allotrope called twin T-graphene that has three atomic layers of thickness was explored by Majidi R. et al. [30]. Twin graphene is created by substituting carbon dimers for one-third of the parallel aromatic bonds in AA-stacked bilayer graphene. It resembles a graphene bilayer in which two layers are connected organically [31]. In their study, density functional theory (DFT) calculations were used to examine the structural and electrical characteristics of 3D twin T-graphene embedded with transition metals (TM). They showed that TM adsorption influences the twin T-electrical graphene's characteristics. The authors observed the adsorption of Sc, Ti, V, Cr, and Zn on semiconductors, Mn, Cu, and Ni on metals, and Fe and Co on bipolar magnetic semiconductors. Twin T-graphene sheets with TM embedded in them have an energy band gap that narrows with increasing TM atom concentration. The outcomes demonstrated the potential of TM-integrated twin T graphene for usage in electrical and spintronic systems. Twin graphene investigations with the dual doping of Al and Y (Y—B, N, O) atoms at various sites (ortho, meta, and para) were also carried out by Yu L. et al. using first-principle DFT calculations to determine the structural, electrical, and magnetic properties [32]. The AlB-TG system is the most stable dual-doped structure since all Al-Y dual-doped twin graphene (AlY-TG) systems are formed through exothermic methods that result in stable dual-doped structures. In the situations of AlB and AlN doping, dual doping controls the bandgap of twin graphene. The bandgap of twin graphene is controlled by doping when AlB and AlN are utilized, and the pure twin graphene has a direct bandgap, is nonmagnetic, and is semiconductive. By dual doping Al and Y (Y—B, N, O) atoms, the twin graphene's electrical and magnetic characteristics can be altered. A theoretical foundation for using twin graphene in nanomagnets and spintronic devices is provided in this paper. Graphene nanoribbons (GNRs) can become the fundamental components of spintronic devices thanks to the embedded spin chains. Magnetism in GNRs is typically linked to highly reactive, hard-to-produce localized states along zigzag edges. Friedrich N. et al. showed that by creating atomically precise engineering topological flaws in the interior of the GNR, magnetism can also be induced away from the physical zigzag edges [33]. Two spin-polarized boundary states are created around a pair of substitutional boron atoms that are introduced into the carbon backbone, breaking the conjugation of their topological bands. Electrical transport experiments using boron-substituted GNRs suspended between the tip and the sample of a scanning tunnelling microscope revealed the spin state. A magnetic ground state is made possible by adding two boron atoms to the carbon lattice of graphene nanoribbons (GNRs) as shown in Figure 4.

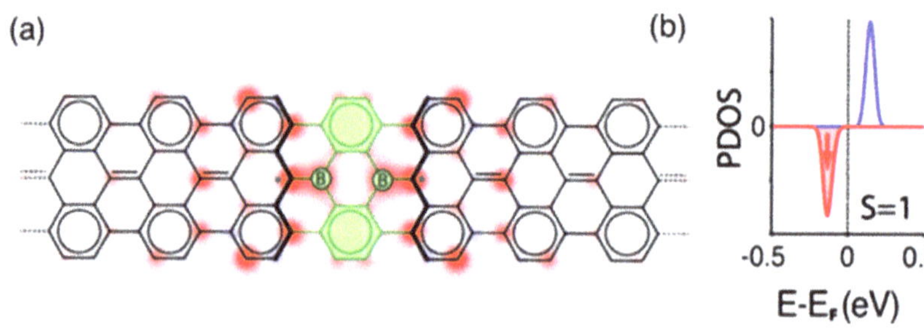

Figure 4. (**a**) Structure of the 2B-7AGNR shown over a color map representing the spin polarization density map, computed using density functional theory simulations ([16]) (green represents the boron

moiety). (**b**) Spin-resolved projected density of states (PDOS) over carbon atoms around the boron dimer. Net spin polarization of one kind confirms the ferromagnetic alignment of the two magnetic moments. Reproduced from ref [33] with permission of Phys. Rev. Lett. copyright 2020. The production of nitrogen-doped porous graphene nanoribbons (N-GNRs) on Ag(111) was accomplished by Pawlak R. in a different study by using a silver-assisted Ullmann polymerization of brominated tetrabenzophenazine [34]. Combining scanning tunneling microscopy (STM), atomic force microscopy (AFM) with CO-tip, scanning tunneling spectroscopy (STS), and density functional theory provides insights into the hierarchical reaction pathways from single molecules toward the formation of one-dimensional organometallic complexes and N-GNRs (DFT).

Osman W. and their colleagues use density functional theory to examine the electrical and magnetic properties of armchair-hexagonal (AHEX) and zigzag-triangular (ZTRI) graphene quantum dots doped with alkali metals [35]. Alkali metals (Li, Na, and K) act as dopants by substituting one of the C-atoms present inside the flake in various locations. Also, the stability of the undoped systems is confirmed using the binding energy. Even though doping makes single-layer structures less stable, bilayer structures have higher binding energy between the layers. Magnetic characteristics are also influenced by stacking; for example, edge pairing causes pure bilayer triangular flakes to turn antiferromagnetic. Doping has a considerable impact on the energy gap; for example, when Na is doped at the upper position in hexagonal flakes, the gap drops from 3.7 eV to 1.5 eV. Zheng Y. et al. showed that designer above-room-temperature magnetic phases and functionalities are abundant in graphene nanomaterials. The capacity to adjust magnetic coupling signs has remained elusive but highly wanted, despite recent confirmation that spins exist in open-shell nanographenes. In atomically accurate open-shell bipartite/non-bipartite nanographene, they have successfully illustrated an engineering method for magnetic ground states by combining scanning probe techniques with mean-field Hubbard model calculations [36]. By disrupting the bipartite lattice symmetry of nanographene, the magnetic coupling direction between two spins was regulated. Additionally, by carefully adjusting the overlap of two spins' spin densities, the exchange interaction strength between them was broadly controlled, yielding a significant exchange interaction strength of 42 meV.

3.1.2. Logic Application:

Similar to the implantation of biological logic-gates [37–39], graphene nanoflakes (GNFs) have recently gained significant interest due to their use in spintronic devices [40,41]. They allow for the induction of magnetism via boundary states, defects, doped magnetic atoms, or strain. In a current study, the authors suggested a number of binary (two-qubit) logic gates in the p-conjugated rhombic graphene nanoflakes (Co_4-GNF). For the Co_4-GNF structure, the spin manipulation procedure was completed with a fidelity above 96% in the subpicosecond time scale. They used the position and direction of the spin as the information bits of the binary gates because the spin density of the system continues to be highly localized in some electrical states. Numerous options result in various reorganizations of various classical and quantum logic gates. Both traditional (OR, AND, NAND) and quantum (CNOT, SWAP) binary logic gates are built by carefully mixing the various spin-dynamics processes attained in the Co_4-GNF structure [27]. A molecular spin logic gate was designed by Zhang W. et al. using two Mn porphyrins connected by a six-carbon monoatomic chain (diMnPh) and sandwiched between two electrodes made of armchair graphene nanoribbons (AGNR). The researchers' findings demonstrate that the spin-resolved transport features can be successfully controlled by simultaneously modifying the molecules' initial spin polarizations and zigzag edges. The diMnPh molecular junction can realize various spin logic gates, such as YES, NOT, XOR, OR, and NOR, which are crucial for designing and implementing high-performance and multifunctional molecular spintronic devices in the future. These gates can be realized based on different spin-dependent current-voltage characteristics [42]. The work by the Han Zhou group thoroughly examines the spin couplings between transition metal atoms doped on graphene and demonstrates how they may be used to create various logic gates for spintronic device design [43]. The spin-coupling effect can manifest a certain distance dependence and space propagation, as further confirmed by the impacts of the number of carbon layers and the distance between doped metal atoms on the logic gate implementation. The accomplishments in this work reveal the potential utility of graphene materials and are anticipated to open new research directions for investigating their use in the creation of complex spintronic devices. In this work, the six logic gates AND, OR, NOT, XOR, NOR, and INHIBIT are implemented.

3.2. Carbon Nanotubes

Since Iijima's 1991 discovery of CNTs, there has been a lot of interest in CNT-based spintronics [12]. Different structural categories are used to categorize CNTs. Single-walled carbon nanotubes (SWCNTs) are created by rolling a graphene nanoribbon into a cylinder; multi-walled carbon nanotubes are created by nesting several SWCNTs with various radii together (MWCNTs). Depending on their diameters and chiralities, SWCNTs can have metallic or semiconducting electrical characteristics. MWCNTs can be thought of as a collection of coaxial SWCNTs.

3.2.1. SWCNTs

The design of magnetic materials, spintronic devices, and others are all inextricably connected to the spin-polarized behavior of electrons [3,44]. The research of wang J. et al. on cap-(9, 0) CNTs investigates a defect for adsorbed atoms that could lead to spin polarization on the surface of C30 [45]. To investigate it, they employed the first-principles DFT approach [46–48]. The outcomes of the calculations demonstrate that the C adatom drives the formation of spin density and that the distribution of spin density varies for various adsorption positions. The research on spintronic injection and other devices will benefit from this effort. Their research suggests that the asymmetric structure's spin polarization can be modified by adding adatom defects. The capacity of spin crossover (SCO) molecules to change their spin state in response to various stimuli makes them interesting candidates for nanoscale magnetic switches. Strong Fe-based SCO molecules have proven to be enclosed within the 1D cavities of single-walled carbon nanotubes by Villalva J. et al. [49] as shown in Figure 5. They discovered the SCO process holds up to individual heterostructures being enclosed and placed in nanoscale transistors. A substantial conductance bistability is triggered through the host SWCNT via the SCO switch in the guest molecules. Additionally, unlike crystalline samples, the SCO transition occurs at higher temperatures and exhibits hysteresis cycles, which results in the memory effect. Their findings show how SCO molecules can be processed and positioned into nanodevices using SWCNTs as a support structure, which can also help to adjust their magnetic characteristics. Using DNA strands to helical functionalize carbon nanotubes, the chirality-induced spin selectivity (CISS) effect can polarize carrier spins.

A fundamental comprehension of this effect, which M. W. Rahman and colleagues have studied, is essential for the prospective application of this system in spintronic devices [50]. Due to DNA functionalization, the conduction mechanism was discovered to operate in the strongly localized regime, and the observed magnetoresistance is a result of interference between the forward and backward hopping paths. According to estimates, CISS-induced spin polarization increases the carrier localization length by an order of magnitude in the low temperature range and has a non-trivial impact on the magnetoresistance effect that was not seen in traditional systems.

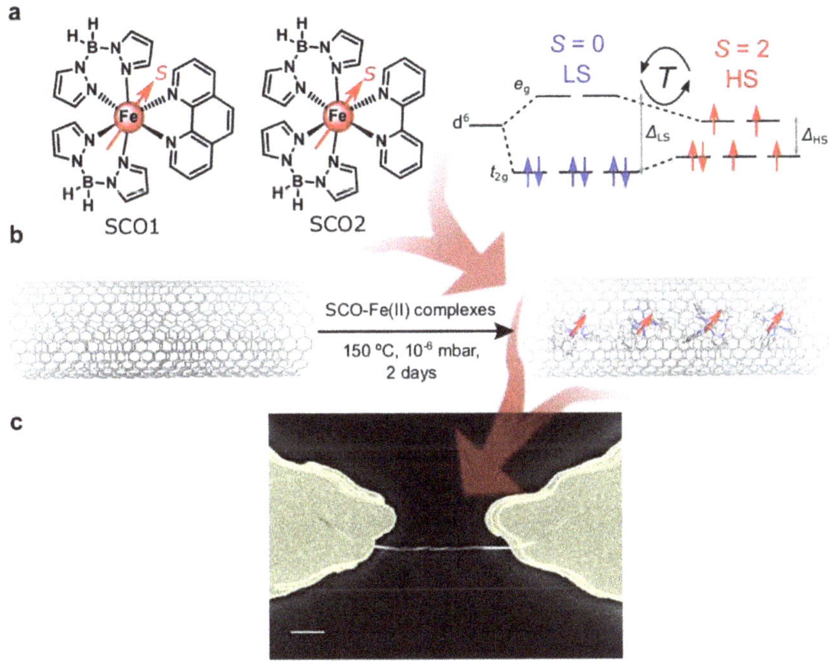

Figure 5. (**a**) Complex SCO1, [Fe(H_2Bpz_2)$_2$(phen)] and SCO2, [Fe(H_2Bpz_2)$_2$(bipy)] encapsulated in this work. Spin (S) level distribution in the high (HS) and low (LS) spin states of the molecules. (**b**) Schematic pathway for the encapsulation of Fe (II) SCO complexes in oSWCNTs, resulting in SCO@SWCNT. (**c**) Scanning electron microscopy (SEM) image of a transistor-like device containing a SCO2@SWCNT heterostructure trapped via dielectrophoresis. Scale bar: 500 nm. Reproduced from ref [49] with permission of Nat. Commun. copyright 2021.

To create high-performance nanoelectronic devices in the future with minimal interface contact resistance barriers and micro superconducting spintronic devices, C/BN heteronanotubes with polar discontinuity have been created by Wang Y. [51]. Linear interfaces between CNTs and h-BN nanotubes (BNNTs) within heteronanotubes introduce unexpected electrical and spintronic properties that are distinct from those of the individual parts. Polar discontinuities occur at the interfaces of heteronanotubes with zigzag boundaries (ZCBNNTs), creating an inherent electric field that acts on the segment CNTs. To make up for the potential difference, the electric field then induces a charge transfer between the two interfaces; as a result, the interface states (or boundary states) exhibit novel behaviors. For instance, ZCBNNTs are half-metallic in nature. Due to the compensatory effects of the free charge to the polarization charge, the ZCBNNTs' length and diameter can also be used to alter the electrical properties in an efficient manner. Chen M. et al. developed innovative nanomaterials for use in electrical atomic switches, optoelectronic, and spintronic devices [52]. They have shown that chromium atoms put between parallel SWNT sidewalls can transfer electrons to the benzene rings of the nanotubes by forming hexahapto bonds with the benzene atoms. This maintains the nanotube's conjugated electrical structure. An appealing method for the reversible chemical engineering of the transport characteristics of aligned carbon nanotube thin films is to link the graphitic surfaces of carbon nanotubes with transition metal atoms, which boosts the transverse conductivity of connected and aligned SWNTs by ≈2100%. They show that a SWNT-aligned device can transition back and forth between a state of high electrical conductivity. They have shown that a device made of aligned SWNTs is capable of being reversibly switched from a state of high electrical conductivity (ON) via light to a state of low electrical conductivity (OFF) by applying voltage. CNTs are excellent choices for supercapacitor electrodes, both when combined with other electrode materials and when used alone. Activating the CNT walls and/or tips will increase the specific surface area. For instance, Pan et al. increased the specific surface area of SWNTs through electrochemical activation from 46.8 m^2/g to 109.4 m^2/g, resulting in a three-fold increase in the specific capacitance [53]. For extremely pure SWNTs, Hata and colleagues reported a specific surface area of 1300 m^2/g [54]. Chen X. et al. have reported energy densities up to 94 Wh/kg (or 47 Wh/L) and power densities up to 210 kW/kg (or 105 kW/L) using an organic electrolyte (1 M Et4NBF4/propylene carbonate) to assure a high voltage of 4 V [55].

3.2.2. MWCNTs

Danilyuk A.L. and colleagues have explored the indirect exchange coupling mediated by conduction electrons in multiwall carbon nanotubes (MWCNTs) with single-domain ferromagnetic nanoparticles (FNPs) implanted inside. They have demonstrated that the static spin susceptibility can spread up to tens of micrometers by modifying the Klinovaja–Loss (KL) model for single-wall CNTs [56]. The adjustment of the Fermi level to the gap created by the spin–orbit interaction is the primary requirement for the long-range exchange interaction (SOI). The suggested method enables measuring the exchange interaction's energy between FNPs that are part of the same CNT. The results obtained present promising prospects for the fabrication and application of MWCNT-based spintronic devices. Mosse I.S.'s work with nano-tweezers is an important source for the creation of device components that could be useful for quantum information technology [57]. A controlled synthetic chemical technique was used to improve magnetic interactions along nanotube walls. This technique is based on a two-step process that first examines the functionalization of nanotubes (carbonyl groups) and then looks at the attachment of an organo–metallic complex to the carbonyl group. Depending on the functionalization method, mesoscopic electron spin correlations have been seen as well as a distinct transition from superparamagnetism to weakly ferromagnetism. Next, they used a nano-tweezer made from a memory metal alloy to illustrate a novel production method based on nanointegration. The developed devices had quantum rings, crossed junctions, and fine network topologies that can be controlled by nano-probes.

3.3. *Carbon Nitride*

3.3.1. General

A brand-new group of carbon-based materials made up of nitrogen (N) and carbon (C) atoms are known as 2D carbon nitride nanosheets. A recent study has focused heavily on the 2D carbon nitride family due to evidence of its physical, chemical, and morphological features [58,59]. It is interesting to note that by adjusting the N/C ratio and moving the N and C atoms around in the lattice structure, the electric characteristics of carbon nitride nanosheets can be flexibly designed to range from half-metal to semiconductor [60,61], offering insightful design guidance for novel g-C3N4-based two-dimensional elastic electrical and spintronic devices. Biaxially strained graphitic carbon nitride's (g-C3N4) mechanical and electrical characteristics were studied by Qu L.-H. and colleagues [62]. The results demonstrated that g-C3N4 has substantial linear elasticity and extremely isotropic mechanical characteristics. The photon transition between band gaps was modest, implying that the g-C3N4 monolayer is not a good material for solar cells. It was discovered that the spin-unrestricted band gap of g-C3N4 can be over-estimated, and that sufficient biaxial strain can cause the spin splitting of g-C3N4. The experimental implementation of bare C3N nanoribbons spintronic devices is challenging. This is because edge reconstruction will take place in the bare C3N nanoribbons. Utilizing ferromagnetic Ni electrodes, Zeng J. and Zhou Y. thoroughly examined the spin-polarized characteristics of 2D C3N sheets and the related hydrogen-passivated nanoribbons [63]. On top of the Ni electrodes, a 2D C3N sheet exhibited semiconductor-to-metallic spin filtering and a positive magnetoresistance effect. Particularly impressive were the increased spin polarization effectiveness and the appearance of the negative magnetoresistance effect following the transformation of the 2D C3N sheet into hydrogen-passivated nanoribbons. These findings suggest that C3N has a lot of promise for use in nanoscale spintronics.

3.3.2. Doping

In addition to using carbon nitride in its purest form, doping the monolayer with atomic impurities has also been investigated to enhance its physical and chemical properties. By manipulating the band structure, doping with non-metal atoms such as S, O, B, and C can increase the lifetime of photogenerated electrons and the range of absorbed light wavelengths [64]. One efficient and convenient approach to improvement is co-doping carbon nitride nanosheets. According to research by Bafekry et al., co-doping changes the semiconducting properties of C6N7 into a Dirac half-metal with a 2.3 eV band gap in the spin-up channel, and a gapless Dirac conic band structure in the spin-down channel. These results show that the 2D carbon nitride family has applicability in spintronic devices, made possible by the co-doping technique, in the band structure [61]. The metal-adsorbed C4N systems are remarkable potential candidates for the development of electrical and spintronic devices. Liu W. and colleagues presented a novel possibility for controlling the ferromagnetism in light element systems [65], by adding ferromagnetism to materials with only s and p electrons in light elements. In this study, the ferromagnetic properties of carbon-doped boron nitride (B-C-N) nanosheets were investigated. These nanosheets were produced by the high-temperature annealing of a stacked mixture of boron nitride nanosheets (BNNSs) and graphene that had been prepared by urea-assisted aqueous exfoliation. The BCN nanosheets showed strong ferromagnetic responses, with a saturation magnetization of 0.142 emu/g at ambient temperature. They discovered that the ferromagnetic characteristics of the BCN nanosheets are significantly influenced by the thickness of the predecessors.

The electrical and magnetic characteristics of the 2D materials were successfully altered by surface adsorption. Xu M. et al. used the density functional theory to examine the adsorption behaviors of 16 metal atoms that were adsorbed on the 2D dumbbell C4N (DB C4N), including alkali metals, alkaline earth metals, 3D transition metals (TMs), and precious metals [66]. The alkali metals (Li, Na), alkaline-earth metals (Be, Mg), 3dTMs (Ti, V, Cr, Mn, Fe, Co), and precious metals (Ag, Pt, Au) were shown to open the zero bandgaps of DB C4N through the charge transfer between the adatoms and the surface of the material. Additionally, the monolayer C4N was made magnetic using the 3D transition and precious metal adatoms.

3.4. *Fullerenes*

We must be able to comprehend and manipulate metal-organic interactions for the functionalization of organic complexes for next-generation electrical and spintronic devices. The so-called single molecular magnets (SMM) are of special relevance for magnetic data storage applications because they provide the opportunity to store information on a molecular scale. The research of Seidel J. et al. focused on the adsorption characteristics of the archetypal SMM Sc3N@C80 produced as a monolayer film on the Ag(111) substrate [67]. They offered convincing proof that the adsorption on the Ag(111) surface caused a pyramidal deformation of the otherwise planar Sc3N core inside the carbon cage. It is possible to link this adsorption-induced structural change in the Sc3N@C80 molecule to a charge transfer from the substrate into the lowest unoccupied molecular orbital of Sc3N@C80, which drastically changes the charge density of the fullerene core. This research demonstrated how such an indirect interaction between the metal centers of SMMs that are enclosed and the metal surfaces can significantly modify the geometric shape of the metallic centers, potentially changing the magnetic characteristics of SMMs on surfaces as well.

3.5. Others

Carbon-based materials have received considerable and widespread research attention ever since monolayer graphene was successfully synthesized. They show great promise for use in electronic devices, even replacing silicon-based electronics, optoelectronics, and spintronics, due to their outstanding transport capacity and conductivity. In work by Liu C. et al., amorphous carbon nanosheets (CNs) with an average of 3.6 nm graphene nanocrystals inside were found to macroscopically attain room-temperature ferromagnetic ordering [11]. The magnetization that could result from the ferromagnetic coupling between the zigzag edges of nanocrystals in CNs would be substantial (>0.22 emu/g), which is two orders of magnitude more than what has been observed in defective graphite. Additional experiments and first-principles calculations showed that the zigzag edges' separation might effectively influence the magnetic coupling in CNs.

In another study, M.A.M. Keshtan and M. Esmaeilzadeh investigated the topological and spin-dependent electron transport characteristics of a trans-polyacetylene molecule [68]. Even though their Hamiltonians do not adhere to chiral symmetry, it was discovered that molecules with intracellular single carbon–carbon bonds and an even number of monomers in their chains exhibit two edge states and contain topological features. The quantum spin-dependent electron transport features are induced and manipulated using two perpendicular and transverse electric fields and a perpendicular exchange magnetic field. In distinct electron energy zones that are extended by stronger exchange fields, the exchange field causes spin polarization. As a result, the suggested gadget functions perfectly as a spin filter. A perfect spin caloritronics device with a carbon-based organic chain was proposed by Tan F. et al. It is possible to achieve a spin-semiconducting feature that results from edge localized states [69]. A significant spin Seebeck coefficient emerges from the spin-dependent transport gaps. Moreover, at room temperature, the dimensionless spin thermoelectric figure of merit (FOM) can be increased to 35. As a result of a temperature difference, it is also possible to produce a pure spin current or single-spin current at some chemical potentials, and the chemical potential can also be used to control the transport directions of these currents. Optically addressable spins, which combine a long-lived qubit with a spin-optical interface for external qubit control and readout, are a promising framework for quantum information science. A modular qubit architecture is made possible by the ability to chemically synthesize such systems, which can produce optically addressable molecular spins. This architecture can be transported between different environments and atomistically customized for specific applications through a bottom-up design and synthesis. D. D. Awschalom and colleagues show how manipulating the host environment can affect the spin coherence in such optically addressable molecular qubits [70]. They create noise-insensitive clock transitions utilizing chromium (IV)-based molecular qubits in a nonisostructural host matrix through a transverse zero-field splitting that is not possible with an isostructural host. In a nuclear and electron spin-rich environment, this host-matrix engineering produces spin-coherence times of more than 1.0 μs for optically addressable molecular spin qubits. Their findings show how a customizable molecular platform may be used to evaluate qubit structure–function relationships and highlight potential applications for employing molecular qubits for nanoscale quantum sensing in noisy environments.

3.6. Doped and Codoped

The properties of carbon materials can be enhanced or modified through the use of adatoms, adsorption, doping, defects, the addition of an electric field, changes in tension, and other techniques. Based on the density functional theory, Zhongyao Li and Min Chen investigated the potential half-metallic behavior in three-dimensional, transition metal (Fe, Co, and Ni)-decorated two-dimensional polyaniline (C3N). The Ni-decorated polyaniline ((C3N)2Ni) is a nonmagnetic semiconductor with an increased band gap, according to the estimated electronic structures, but the Fe and Co decorated polyanilines ((C3N)2Fe and (C3N)2Co) are magnetic half-metals. The spacing between 3d transition metal atoms and C3N can change the energy windows and band gaps. 3d-transition-metal-adorned C3N can be used in nanoscale spintronic devices because of its wide half-metallic energy window and suitable band gap. Xia B. et al. established a method for introducing magnetism into carbon nanosheets using a single Cr cation that is only attached to two-dimensional carbon nanosheets via Cr-N bonds [71]. The highest magnetization (Cr: 2.0%, 0.86 emu g^{-1}) under 3 T was achieved at 50 K, where the magnetization changes with the Cr concentration. In the samples, it was shown that the anchoring of Cr can cause paramagnetism and ferromagnetism, and that the magnetization is strongly correlated with the Cr level. A method to create a magnetic carbon matrix was presented in this study, which lays the groundwork for several potential spintronics device applications to carbon-based, low-dimensional materials. A third-generation semiconductor material with a wide band, silicon carbide (SiC), is important in the power and electronics industries. SiC has more than 200 distinct polytypes or crystal lattice alteration patterns. One of the significant polytypes [72] that has received a lot of attention is 4H-SiC. Long Lin's group has thoroughly studied the electronic structure, magnetic, and optical properties of 4H-SiC doped with a single V atom, single Fe atom, and (V, Fe) co-doped 4H-SiC [73]. The 3d orbitals of V or Fe can contribute significantly to the single V or Fe dopant's ability to introduce magnetism in pure 4H-SiC. The coupling relationship between the V-3d, C-2p, and Fe-3d states is believed to be the cause of ferromagnetism, and the (V, Fe) co-doped system prefers FM states. Red-shift phenomena emerge in the absorption spectrum as a result of the addition of V and Fe atoms, greatly enhancing the absorption strength in visible light. The outcomes demonstrate that the (V, Fe) co-doped 4H-SiC system will offer a potential method for the advancement of spintronic devices and optical applications in the future. Table 1 summarize some important carbon- based material spintronic applications.

Table 1. Summary of carbon-based spintronic application.

Carbon-Based Material	Application	References
Carbon nanosolenoid (CNS)	Riemann surfaces	[15]
Partially hydrogenated graphene (PHGr)	Single atom catalysts	[16]
Five-membered ring-decorated nanographene	Overcurrent protection	[29]
Zigzag-edged graphene nanoribbons (ZGNRs)	Spin-filtering device	[32]
Zigzag-edge graphene nanoribbon (ZGNR) electrodes	Spin filtering and the negative differential resistance (NDR) effect	[35]
Boron atoms (B2) doped 7-atom-wide armchair graphene nanoribbons (B2-7AGNRs)	Pie magnetism and spin-dependent transport	[28]
Graphene lattice with transition metal atom	High magnetic anisotropy energy (MAE)	[17]
Manganese porphyrin molecules connected to graphene electrodes	Multifunctional spin logic gates	[42]
Metal-doped graphene	Logic gate application	[43]
Metal embedded twin T graphene	Bipolar magnetic semiconductor for Fe and Co	[30]
DNA-functionalized carbon nanotubes	Spin filters	[50]
Carbon nanotubes (CNTs)	Spintronic injection	[45]
Carbon nanotubes (CNTs) with embedded ferromagnetic materials	Indirect exchange coupling	[52]
Spin cross over SWCNT	Nanoscale magnetic switches	[49]
Carbon nitride C6 N7 atomic doping	Two-dimensional Dirac half-metal	[61]
C3N	Semiconductor-to-metallic spin filtering	[63]
Carbon-doped boron nitride nanosheets	Manipulation of ferromagnetism in light element systems.	[65]

4. Spin Relaxation in Carbon-Based Materials

The conduction electron spin lattice relaxation time (CESR), T_1, is the typical time for a spin system that has been driven out of equilibrium by, for example, a microwave field at electron-spin resonance ESR or a spin-polarized current to return to thermal equilibrium. A sufficiently long spin lifetime is necessary to be used in "spintronics" devices, in which electron spins are used to process information [74]. In the MgB2 superconductor, Simon F. and colleagues measured the spin lattice relaxation time T_1 of conduction electrons as a function of temperature and magnetic field [75]. During electron-spin resonance conditions with amplitude-modulated microwave stimulation, researchers employ a technique based on the detection of the z component of the conduction electron magnetization. Despite the considerable CESR line broadening caused by irreversible diamagnetism in the polycrystalline sample, a lengthening of T_1 below Tc (critical temperature) is seen. They can measure the individual contributions to T_1 from the two different forms of the Fermi surface because of the field independence of T_1 for 0.32 and 1.27 T. A phase-pure crystal of potassium-doped p-terphenyl, [K(222)]2[p-terphenyl3], is isolated by Gadjieva N.A. et al. [76]. Magnetometry and electron spin resonance (ESR) are used to conduct in-depth research on the emerging antiferromagnetism in the anisotropic structure. The antiferromagnetic coupling in this system, which occurs in all three crystallographic directions, has been described by the authors using these experimental findings in combination with calculations using the density functional theory. The terphenyls' ends, where an extra electron on nearby p-terphenyls antiferromagnetically couples, showed the strongest coupling. These results suggest that potassium-doped p-terphenyl exhibits magnetic fluctuation-induced superconductivity, which is closely related to high Tc cuprate superconductors.

Using liquid ammonia, Markus B.G. et al. reported synthesizing few-layer graphene (FLG) doped with Li and Na. Chemical exfoliation was used to prepare the FLG material [77]. The appearance of powerful, metallic-like electron spin resonance (ESR) modes and the modification of the Raman G line into a Fano line shape are evidence of the high concentration of graphene doping in liquid ammonia for both types of alkali atoms. The spin-relaxation duration in the materials was 6–8 ns, which is equivalent to the longest values discovered in spin transport tests on ultrahigh-mobility graphene flakes. This time was calculated from the ESR line width. This might make this substance a good contender for spintronics devices. However, since sodium is a very common metal, a successful sodium doping attempt might be a promising replacement for lithium batteries.

The study by Ren-Shu Wang et al. provides a complete set of parameters through a thorough investigation on a K_3C_{60} sample that has been well-characterized [78]. K_3C_{60} is a promising three-dimensional superconducting magnet material with the advantage of the rich carbon abundance on the Earth due to the high upper critical field of 33.0 ± 0.5 T obtained from the direct electrical transport measurements, along with the relatively high critical temperature and large critical current density. The examination of all independently acquired parameters points to the peculiar nature of K_3C_{60}'s superconductivity, including contributions from electron correlations and electron–phonon coupling. By using the ESR technique, paramagnetic centers in heterofullerides with the compositions A_2MC_{60} and AM_2C_{60} (A = K, Rb, M = Mg, Be) were studied by Kytin V.G. and his collegues [79]. The ESR signal was discovered to be composed of two lines with distinct temperature dependences in terms of ESR absorption magnitude. This provides proof that there are at least two different types of paramagnetic centers present. The first type's centers exhibit localized spin behavior, whereas conduction electrons can be used to describe the remaining centers.

The electron spin resonance (ESR) signal of undoped and potassium-doped SWCNTs was investigated by Galambos M. et al. [80]. They identify the signals of the conduction electron spin resonance (CESR), the low intensity impurity, and the superparamagnetic background. Only the alkali atom doping causes the latter to be present. They critically evaluate the possibility that the CESR signal could be residual graphitic carbon, which they categorically rule out, in order to identify it. They provide precise values for the signal intensities, related spin concentration, and g factors. The density of states on the SWCNT assembly can be calculated using the CESR signal intensity.

5. Organic Spintronics

Carbon-based, molecular, or polymeric semiconductors are used in organic electronics because they are inexpensive to produce, mechanically adaptable, and, most critically, have practically limitless and chemically tunable electrical and optical properties. The organic light-emitting device (OLED), which is utilized in displays, is the most widely used commercial organic electronic application (notably in smart phones). By combining organic materials into spintronics and spin dependent effects in organic electronics, organic spintronics strives to merge these two fields. According to Dediu et al. devices with ferromagnetic La0.7Sr0.3MnO3 (LSMO) electrodes and a sexithienyl (6T) spacer exhibit room-temperature magnetoresistance [81]. In order to create lateral devices, LSMO films were shaped into electrodes separated by a small gap (100–500 nm). The gadget's resistance had a distinct magnetic field dependence. The initial investigations on organic spin valves were motivated by this work and used ferromagnetic LSMO and Co electrodes in a vertical layer stack (layers deposited on top of each other) separated via an Alq3 spacer (thickness 100–200 nm). Utilizing the unique material characteristics of organic semiconductors for spintronic applications is gaining popularity. In a small molecule system based on dinaphtho[2,3-b:2,3-f]thieno[3,2-b]thiophene (DNTT), Angela Wittmann et al. investigate the application of a pure spin current from Permalloy at ferromagnetic resonance [82]. They are able to systematically examine the influence of interfacial characteristics on the spin injection efficiency via molecular design thanks to the unique tunability of organic materials. We demonstrate that the interfacial molecular structure and side chain substitution of the molecule allow delicate tuning of the spin injection efficiency at the interface as well as the spin diffusion length. Organic photovoltaics (OPVs) based on non-fullerene acceptors have received a lot of interest in the past 10 years because of their excellent potential to achieve high-power conversion efficiencies. The primary difficulties in permitting effective charge separation/transport and a low voltage loss simultaneously are what limit the development of higher performance OPVs. To match the commercially available polymer PM6, Yuan J. et al. have designed and created a new type of non-fullerene acceptor, Y6, that uses an electron-deficient, core-based central fused ring with a benzothiadiazole core [83]. With both conventional and inverted architecture, the Y6-based solar cell achieves a high-power conversion efficiency of 15.7% using this method. By doing this study, we offer fresh perspectives on how to build new non-fullerene acceptors to achieve increased photovoltaic performance in OPVs by utilizing the electron-deficient, core-based central fused ring.

6. Challenges in Carbon-Based Spintronics

Twisteronics is currently a popular issue in the study of 2D materials. Although the superconductivity in twisted graphene systems appears to be of an uncommon type, the underlying theoretical description is still elusive, which poses a significant issue for carbon-based spintronics. Increasing production rate, scalability, homogeneity, and quality control are problems for the industrial-scale synthesis of carbon-based spintronic materials. In general, varying quality across producers is a problem, and it is imperative to improve a single standard or grading system. Improving the quality and reproducibility of patterned nanostructures, effectively coupling light into and out of graphene, and expanding plasmon tunability to the vis-NIR range are challenges in plasmonic and optical qualities. Developing regulated fabrication processes that result in reliable and repeatable devices is another problem in electrical and spintronic applications. Of course, there are also significant challenges with scalability and wafer-scale integration. Nanotube-based electronics have two main difficulties that must be overcome. Connectability is one of the difficulties; while it is one thing to construct a single nanotube transistor, it is quite another to connect millions of them. A single component at a time is often used in the current techniques of nanotube electronics, which is not practical. The suggested theoretical research and experimental findings typically differ significantly, since the experiments are so complex. These disagreements suggest that elaborate physical mechanisms and precise theoretical models must be established to effectively direct research efforts and interpret experimental findings. The limitation in material science with carbon-based spintronics is that the majority of materials with thicknesses close to the atomic level are temperature-, oxygenation-, and moisture-sensitive. They need to operate above room temperature and be stable in the presence of air.

7. Outlook and Future Perspective

Due to their unique spin-dependent characteristics, such as lengthy spin relaxation durations, long diffusion lengths, and high spin–orbit coupling, carbon-based materials have received a lot of attention in the field of spintronics. Because of its high charge carrier mobility, long spin lifetime, and long diffusion length, graphene has distinguished itself as a superior platform for future spintronic devices. Effective spin logic and non-volatile data storage are made possible by the tunable bandgap and robust spin–orbit coupling that carbonitrides show. The integration of spintronics and photonics into a single platform for both light-based and spin-based quantum computing is the goal of ongoing research and development. Researchers hope to be able to control the electron spin dynamics in nanostructured 2D magnetic materials activated using brief laser pulses by employing specially created photonic circuits. The rapidly developing field of spintronics, which has contributions from a wide range of nations and disciplines, including biology, chemistry, physics, electrical engineering, computer science, and mathematical information theory, promises to make fundamental discoveries in both pure and applied science as well as have a significant impact on future technology. In contrast to traditional charge-based information processing technologies, using spin as a state variable in logic devices has various benefits, including non-volatility, faster and more energy-efficient data processing, and higher integration densities. Spintronic computing has the potential to meet the constantly rising performance requirements of upcoming abundant-data applications.

Author Contributions: Writing—original draft preparation, S.P., Writing—Review & Editing, S.P., H.D. and D.F.; Supervision, D.F. All authors have read and agreed to the published version of the manuscript.

Funding: This research received no external funding.

Data Availability Statement: All data are available upon email request. Restrictions apply to the availability of these data. Some data are not publicly available since some articles are not open access.

Conflicts of Interest: The authors declare no financial or commercial conflict of interest of data; in the writing of the manuscript, or in the decision to publish the results.

References

1. Hirohata, A.; Yamada, K.; Nakatani, Y.; Prejbeanu, L.; Diény, B.; Pirro, P.; Hillebrands, B. Review on spintronics: Principles and device applications. *J. Magn. Magn. Mater.* **2020**, *509*, 166711. [CrossRef]
2. Baibich, M.N.; Broto, J.M.; Fert, A.; Van Dau, F.N.; Petroff, F.; Eitenne, P.; Creuzet, G.; Friederich, A.; Chazelas, J. Giant magnetoresistance of (001)Fe/(001)Cr magnetic superlattices. *Phys. Rev. Lett.* **1988**, *61*, 2472–2475. [CrossRef]
3. Wolf, S.A.; Awschalom, D.D.; Buhrman, R.A.; Daughton, J.M.; Von Molnár, S.; Roukes, M.L.; Chtchelkanova, A.Y.; Treger, D.M. Spintronics: A spin-based electronics vision for the future. *Science* **2001**, *294*, 1488–1495. [CrossRef] [PubMed]
4. Chen, P.; Zhang, G. Carbon-based spintronics. *Sci. China Phys. Mech. Astron.* **2013**, *56*, 207–221. [CrossRef]
5. Zhang, Y.; Guo, L.; Zhu, X.; Sun, X. The Application of Organic Semiconductor Materials in Spintronics. *Appl. Org. Semicond. Mater. Spintron. Front. Chem* **2020**, *8*, 589207. [CrossRef] [PubMed]
6. Hu, G.; Xiang, B. Recent Advances in Two-Dimensional Spintronics. *Nanoscale Res. Lett.* **2020**, *15*, 226. [CrossRef]
7. Ahn, E.C. OPEN 2D materials for spintronic devices. *NPJ 2d Mater. Appl.* **2020**, *4*, 17. [CrossRef]
8. Zlobin, I.S.; Nelyubina, Y.V.; Novikov, V.V. Molecular Compounds in Spintronic Devices: An Intricate Marriage of Chemistry and Physics. *Inorg. Chem.* **2022**, *61*, 12919–12930. [CrossRef]
9. Pandey, S.; Karakoti, M.; Surana, K.; Dhapola, P.S.; SanthiBhushan, B.; Ganguly, S.; Singh, P.K.; Abbas, A.; Srivastava, A.; Sahoo, N.G. Graphene nanosheets derived from plastic waste for the application of DSSCs and supercapacitors. *Sci. Rep.* **2021**, *11*, 3916. [CrossRef]

10. Kalluri, A.; Debnath, D.; Dharmadhikari, B.; Patra, P. Graphene Quantum Dots: Synthesis and Applications. *Methods Enzymol.* **2018**, *609*, 335–354.
11. Liu, C.; Yang, Y.; Ma, Z.; Zhou, C.; Liu, D.; Luo, X.; Zhu, X.; Sun, Y.; Sheng, Z. Edge-Induced Room-Temperature Ferromagnetism in Carbon Nanosheets. *J. Phys. Chem. C* **2020**, *124*, 7396–7403. [CrossRef]
12. Helical microtubules of graphitic carbon. *Nature* **1991**, *354*, 56–58.
13. Ali, E.; Hadis, D.; Hamzeh, K.; Mohammad, K.; Nosratollah, Z.; Abolfazl, A.; Mozhgan, A.; Younes, H.; Woo, J.S. Carbon nanotubes: Properties, synthesis, purification, and medical applications. *Nanoscale Res. Lett.* **2014**, *9*, 393.
14. Mittal, D.; Dutta, D.P. Synthesis, structure, and selected photocatalytic applications of graphitic carbon nitride: A review. *J. Mater. Sci. Mater. Electron.* **2021**, *32*, 18512–18543. [CrossRef]
15. Wang, J.; Zhu, Y.; Zhuang, G.; Wu, Y.; Wang, S.; Huang, P.; Sheng, G.; Chen, M.; Yang, S.; Greber, T.; et al. Synthesis of a magnetic π-extended carbon nanosolenoid with Riemann surfaces. *Nat. Commun.* **2022**, *13*, 1239. [CrossRef]
16. Song, Y.; Lei, T.; Zhang, Y.; Shixuan, D.U. A DFT Investigation on the Electronic Structures and Au Adatom Assisted Hydrogenation of Graphene Nanoflake Array. *Chem. Res. Chin. Univ.* **2021**, *37*, 1110–1115. [CrossRef]
17. Zeng, J.; Deng, X. Magnetic coupling engineering-induced excellent spin transport in covalently linked nanographene dimers. *Phys. E Low-Dimens. Syst. Nanostruct.* **2021**, *134*, 114935. [CrossRef]
18. Mishra, S.; Beyer, D.; Eimre, K.; Ortiz, R.; Fernández-Rossier, J.; Berger, R.; Gröning, O.; Pignedoli, C.A.; Fasel, R.; Feng, X.; et al. Collective All-Carbon Magnetism in Triangulene Dimers. *Angew. Chem.—Int. Ed.* **2020**, *59*, 12041–12047. [CrossRef]
19. Keerthi, A.; Sánchez-Sánchez, C.; Deniz, O.; Ruffieux, P.; Schollmeyer, D.; Feng, X.; Narita, A.; Fasel, R.; Müllen, K. On-surface Synthesis of a Chiral Graphene Nanoribbon with Mixed Edge Structure. *Chem.—Asian J.* **2020**, *15*, 3807–3811. [CrossRef]
20. Gu, J.; Hu, J.; Zhang, W. Enhancing spin-orbit coupling in high-mobility graphene by introducing chiral space curvature. *New J. Phys.* **2021**, *23*, 043031. [CrossRef]
21. de Sousa, M.S.M.; Liu, F.; Malard, M.; Qu, F.; Chen, W. Turning Graphene into Nodal-Line Semimetals by Vacancy Engineering. *APS Meet. Abstr.* **2022**, *2022*, 1–7.
22. Hou, L.; Hu, H.; Yang, G.; Ouyang, G. Giant Switching Effect and Spintronic Transport Properties in Cyclo [18]carbon-Based Molecular Devices. *Phys. Status Solidi—Rapid Res. Lett.* **2021**, *15*, 2000582. [CrossRef]
23. Correa, J.D.; Pacheco, M.; Bravo, S.; Chico, L. Electronic and magnetic properties of pentagonal nanoribbons. *Carbon N. Y.* **2020**, *162*, 209–219. [CrossRef]
24. Han, X.; Yang, J.; Yuan, P.; Bian, B. Spin-dependent transport in all-carbon multifunctional spintronic device. *Eur. Phys. J. B* **2019**, *92*, 32. [CrossRef]
25. Mousavi, S.T.; Allah, H.; Gharbavi, K.; Yang, N.; Li, C.; Tang, Y.; Nanodevices, N.; Soliman, W.; Asham, M.D.; Phillips, H. Band gap control of bilayer zigzag graphene nanoribbon by direction of magnetic moment Band gap control of bilayer zigzag graphene nanoribbon by direction of magnetic moment. *J. Phys. Conf. Ser.* **2019**, *1402*, 044106.
26. Zhao, Y.; Hu, C.; Hu, Y.; Cheng, H.; Shi, G.; Qu, L. A versatile, ultralight, nitrogen-doped graphene framework. *Angew. Chem.—Int. Ed.* **2012**, *51*, 11371–11375. [CrossRef]
27. Zhang, Y.; Liu, J.; Jin, W.; Lefkidis, G.; Hübner, W.; Li, C. Two-qubit logic gates based on the ultrafast spin transfer in π-conjugated graphene nanoflakes. *Carbon N. Y.* **2022**, *193*, 195–204. [CrossRef]
28. Zhang, P.; Li, X.; Dong, J.; Zhu, M.; Zheng, F.; Zhang, J. π-magnetism and spin-dependent transport in boron pair doped armchair graphene nanoribbons. *Appl. Phys. Lett.* **2022**, *120*, 132406. [CrossRef]
29. Langer, R.; Mustonen, K.; Markevich, A.; Otyepka, M.; Susi, T.; Błoński, P.B. Graphene Lattices with Embedded Transition-Metal Atoms and Tunable Magnetic Anisotropy Energy: Implications for Spintronic Devices. *ACS Appl. Nano Mater.* **2022**, *2022*, 1562–1573. [CrossRef]
30. Majidi, R.; Ramazani, A.; Rabczuk, T. Electronic properties of transition metal embedded twin T-graphene: A density functional theory study. *Phys. E Low-Dimens. Syst. Nanostruct.* **2021**, *133*, 114806. [CrossRef]
31. Jiang, J.W.; Leng, J.; Li, J.; Guo, Z.; Chang, T.; Guo, X.; Zhang, T. Twin graphene: A novel two-dimensional semiconducting carbon allotrope. *Carbon N. Y.* **2017**, *118*, 370–375. [CrossRef]
32. Yu, B.Y.; Xie, Y.; Wu, X.; Gao, Y.; Wang, S.F.; Zhang, J.M. Structural and electronic properties of AlY (Y[dbnd]B, N, O) dual-doped twin graphene: A density functional theory study. *Phys. E Low-Dimens. Syst. Nanostruct.* **2021**, *128*, 114619. [CrossRef]
33. Friedrich, N.; Brandimarte, P.; Li, J.; Saito, S.; Yamaguchi, S.; Pozo, I.; Penã, D.; Frederiksen, T.; Garcia-Lekue, A.; Sánchez-Portal, D.; et al. Magnetism of Topological Boundary States Induced by Boron Substitution in Graphene Nanoribbons. *Phys. Rev. Lett.* **2020**, *125*, 146801. [CrossRef] [PubMed]
34. Pawlak, R.; Liu, X.; Ninova, S.; D'Astolfo, P.; Drechsel, C.; Sangtarash, S.; Häner, R.; Decurtins, S.; Sadeghi, H.; Lambert, C.J.; et al. Bottom-up Synthesis of Nitrogen-Doped Porous Graphene Nanoribbons. *J. Am. Chem. Soc.* **2020**, *142*, 12568–12573. [CrossRef] [PubMed]
35. Osman, W.; Abdelsalam, H.; Ali, M.; Teleb, N.H.; Yahia, I.S.; Ibrahim, M.A.; Zhang, Q. Electronic and magnetic properties of graphene quantum dots doped with alkali metals. *J. Mater. Res. Technol.* **2021**, *11*, 1517–1533. [CrossRef]
36. Zheng, Y.; Li, C.; Xu, C.; Beyer, D.; Yue, X.; Zhao, Y.; Wang, G.; Guan, D.; Li, Y.; Zheng, H.; et al. Designer spin order in diradical nanographenes. *Nat. Commun.* **2020**, *11*, 6076. [CrossRef]
37. Pawar, S.; Duadi, H.; Fleger, Y.; Fixler, D. Carbon Dots-Based Logic Gates. *Nanomaterials* **2021**, *11*, 232. [CrossRef]
38. Barnoy, E.A.; Popovtzer, R.; Fixler, D. Fluorescence for biological logic gates. *J. Biophotonics* **2020**, *13*, e202000158. [CrossRef]

39. Pawar, S.; Duadi, H.; Fleger, Y.; Fixler, D. Design and Use of a Gold Nanoparticle-Carbon Dot Hybrid for a FLIM-Based IMPLICATION Nano Logic Gate. *ACS Omega* **2022**, *7*, 22818–22824. [CrossRef]
40. Gyanchandani, N.; Pawar, S.; Maheshwary, P.; Nemade, K. Preparation of spintronically active ferromagnetic contacts based on Fe, Co and Ni Graphene nanosheets for Spin-Field Effect Transistor. *Mater. Sci. Eng. B* **2020**, *261*, 114772. [CrossRef]
41. Zhang, W.X.; Shi, C.H.; He, C.; Bai, M. External-strain induced transition from Schottky to ohmic contact in Graphene/InS and Graphene/Janus In2SSe heterostructures. *J. Solid State Chem.* **2020**, *289*, 121511. [CrossRef]
42. Zhang, W.; Zhang, G.-P.; Li, Z.-L.; Fu, X.-X.; Wang, C.-K.; Wang, M. Design of multifunctional spin logic gates based on manganese porphyrin molecules connected to graphene electrodes. *Phys. Chem. Chem. Phys.* **2022**, *24*, 1849. [CrossRef] [PubMed]
43. Zhou, H.; Hu, X.; Fang, W.; Su, N.Q. Revealing Intrinsic Spin Coupling in Transition Metal-Doped Graphene. *Phys. Chem. Chem. Phys.* **2022**, *24*, 16300–16309. [CrossRef]
44. Kuemmeth, F.; Churchill, H.O.H.; Herring, P.K.; Marcus, C.M. Carbon nanotubes for coherent spintronics. *Mater. Today* **2010**, *13*, 18–26. [CrossRef]
45. Wang, J.; Liu, X.; Wang, C.; Zhang, W.; Qin, Z. Adatom Defect Induced Spin Polarization of Asymmetric Structures. *ChemistryOpen* **2022**, *11*, e202100208. [CrossRef] [PubMed]
46. Gallo, M.; Favila, A.; Glossman-Mitnik, D. DFT studies of functionalized carbon nanotubes and fullerenes as nanovectors for drug delivery of antitubercular compounds. *Chem. Phys. Lett.* **2007**, *447*, 105–109. [CrossRef]
47. Lu, X.; Sun, C.; Li, F.; Cheng, H.M. Selected absorption behavior of sulfur on single-walled carbon nanotubes by DFT. *Chem. Phys. Lett.* **2008**, *454*, 305–309. [CrossRef]
48. Parr, R.G. Density Functional Theory. *Annu. Rev. Phys. Chem.* **2003**, *34*, 631–656. [CrossRef]
49. Villalva, J.; Develioglu, A.; Montenegro-Pohlhammer, N.; Sánchez-de-Armas, R.; Gamonal, A.; Rial, E.; García-Hernández, M.; Ruiz-Gonzalez, L.; Costa, J.S.; Calzado, C.J.; et al. Spin-state-dependent electrical conductivity in single-walled carbon nanotubes encapsulating spin-crossover molecules. *Nat. Commun.* **2021**, *12*, 1–8. [CrossRef]
50. Rahman, M.W.; Firouzeh, S.; Pramanik, S. Carrier localization and magnetoresistance in DNA-functionalized carbon nanotubes. *Nanotechnology* **2021**, *32*, 455001. [CrossRef]
51. Wang, Y.; Wei, W.; Wang, H.; Huang, B.; Dai, Y. The diverse electronic properties of C/BN heteronanotubes with polar discontinuity. *J. Phys. D Appl. Phys.* **2019**, *52*, 215302. [CrossRef]
52. Chen, M.; Li, W.; da Silveira Venzel, T.E.; Li, G.; Itkis, M.E.; Haddon, R.C.; Bekyarova, E. Effect of constructive rehybridization on transverse conductivity of aligned single-walled carbon nanotube films. *Mater. Today* **2018**, *21*, 937–943. [CrossRef]
53. Pan, H.; Poh, C.K.; Feng, Y.P.; Lin, J. Supercapacitor electrodes from tubes-in-tube carbon nanostructures. *Chem. Mater.* **2007**, *19*, 6120–6125. [CrossRef]
54. Izadi-Najafabadi, A.; Yasuda, S.; Kobashi, K.; Yamada, T.; Futaba, D.N.; Hatori, H.; Yumura, M.; Iijima, S.; Hata, K. Extracting the Full Potential of Single-Walled Carbon Nanotubes as Durable Supercapacitor Electrodes Operable at 4 V with High Power and Energy Density. *Adv. Mater.* **2010**, *22*, E235–E241. [CrossRef]
55. Chen, X.; Paul, R.; Dai, L. Carbon-based supercapacitors for efficient energy storage. *Natl. Sci. Rev.* **2017**, *4*, 453–489. [CrossRef]
56. Klinovaja, J.; Loss, D. RKKY interaction in carbon nanotubes and graphene nanoribbons. *Phys. Rev. B* **2013**, *87*, 45422. [CrossRef]
57. Mosse, I.S.; De Sousa, A.S.; Ncube, S.; Coleman, C.; Bhattacharyya, S.; Irzhak, A.; Gratowski, S.; Koledov, V. Bottom-up nano-integration route for modified carbon nanotube spintronic device fabrication. *J. Phys. Conf. Ser.* **2020**, *1461*, 012015. [CrossRef]
58. Bafekry, A.; Stampfl, C.; Farjami Shayesteh, S. A First-Principles Study of C3N Nanostructures: Control and Engineering of the Electronic and Magnetic Properties of Nanosheets, Tubes and Ribbons. *ChemPhysChem* **2020**, *21*, 164–174. [CrossRef]
59. Bafekry, A.; Farjami Shayesteh, S.; Peeters, F.M. Introducing novel electronic and magnetic properties in C3N nanosheets by defect engineering and atom substitution. *Phys. Chem. Chem. Phys.* **2019**, *21*, 21070–21083. [CrossRef]
60. Bafekry, A.; Stampfl, C.; Ghergherehchi, M.; Farjami Shayesteh, S. A first-principles study of the effects of atom impurities, defects, strain, electric field and layer thickness on the electronic and magnetic properties of the C2N nanosheet. *Carbon N. Y.* **2020**, *157*, 371–384. [CrossRef]
61. Bafekry, A.; Neek-Amal, M.; Peeters, F.M. Two-dimensional graphitic carbon nitrides: Strain-tunable ferromagnetic ordering. *Phys. Rev. B* **2020**, *101*, 165407. [CrossRef]
62. Qu, L.; Deng, Z.; Yu, J.; Lu, X.; Zhong, C.; Zhou, P. Mechanical and electronic properties of graphitic carbon nitride (g-C 3 N 4) under biaxial strain. *Vacuum* **2020**, *176*, 109358. [CrossRef]
63. Zeng, J.; Zhou, Y. Physica E: Low-dimensional Systems and Nanostructures Semiconductor-to-metallic spin-filtering and positive and negative magnetoresistance effects in C 3 N with nickel electrodes. *Phys. E Low-Dimens. Syst. Nanostruct.* **2020**, *118*, 113861. [CrossRef]
64. Li, J.-C.; Hou, P.-X.; Liu, C.; Li, J.-C.; Hou, P.-X.; Liu, C. Heteroatom-Doped Carbon Nanotube and Graphene-Based Electrocatalysts for Oxygen Reduction Reaction. *Small* **2017**, *13*, 1702002. [CrossRef] [PubMed]
65. Liu, W.; Yanase, T.; Nagahama, T.; Shimada, T. Synthesis of carbon-doped boron nitride nanosheets and enhancement of their room-temperature ferromagnetic properties. *J. Alloys Compd.* **2019**, *792*, 1206–1212. [CrossRef]
66. Xu, M.; Wang, H.; Sun, S.; Li, H.; Li, X.; Chen, Y.; Ni, Y. First-Principles Study of Metal Atoms Adsorption on 2D Dumbbell C4N. *Phys. Status Solidi Basic Res.* **2020**, *257*, 1–8.

67. Seidel, J.; Kelly, L.L.; Franke, M.; Van Straaten, G.; Kumpf, C.; Cinchetti, M.; Aeschlimann, M.; Stadtmüller, B. Adsorption-induced pyramidal distortion of the trimetallic nitride core inside the endohedral fullerene Sc3 N@ C80 on the Ag(111) surface. *Phys. Rev. B* **2018**, *98*, 085434. [CrossRef]
68. Keshtan, M.A.M.; Esmaeilzadeh, M. Topological spintronics in a polyacetylene. *J. Phys. Condens. Matter* **2020**, *32*, 345302. [CrossRef]
69. Tan, F.X.; Yang, L.Y.; Yang, X.F.; Liu, Y.S. Thermoelectric transport properties of magnetic carbon-based organic chains. *Chem. Phys.* **2020**, *528*, 110524. [CrossRef]
70. Bayliss, S.L.; Deb, P.; Laorenza, D.W.; Onizhuk, M.; Galli, G.; Freedman, D.E.; Awschalom, D.D. Enhancing Spin Coherence in Optically Addressable Molecular Qubits through Host-Matrix Control. *Phys. Rev. X* **2022**, *12*, 031028. [CrossRef]
71. Xia, B.; Zhang, H.; Liao, Z.; Wu, J.F.; Hu, Y.; Shakouri, M.; Gao, D.; Xue, D. Cr cation-anchored carbon nanosheets: Synthesis, paramagnetism and ferromagnetism. *Nanotechnology* **2021**, *32*, 335706. [CrossRef] [PubMed]
72. He, M.; He, X.; Lin, L.; Song, B.; Zhang, Z.H. Study on spin polarization of non-magnetic atom in diluted magnetic semiconductor: The case of Al-doped 4H–SiC. *Solid State Commun.* **2014**, *197*, 44–48. [CrossRef]
73. Lin, L.; Yan, L.; Huang, J.; Tao, H.; Zhang, J.; Yu, W. First-principles investigations of the e ff ect of V and Fe dopants on the magnetic and optical properties of 4H-SiC. *Thin Solid Films* **2020**, *709*, 138182. [CrossRef]
74. Žutić, I.; Fabian, J.; Das Sarma, S. Spintronics: Fundamentals and applications. *Rev. Mod. Phys.* **2004**, *76*, 323–410. [CrossRef]
75. Simon, F.; Murányi, F.; Fehér, T.; Jánossy, A.; Forró, L.; Petrovic, C.; Bud'ko, S.L.; Canfield, P.C. Spin-lattice relaxation time of conduction electrons in Mg B2. *Phys. Rev. B—Condens. Matter Mater. Phys.* **2007**, *76*, 024519. [CrossRef]
76. Gadjieva, N.A.; Szirmai, P.; Sági, O.; Alemany, P.; Bartholomew, A.K.; Stone, I.; Conejeros, S.; Paley, D.W.; Hernández Sánchez, R.; Fowler, B.; et al. Intermolecular Resonance Correlates Electron Pairs down a Supermolecular Chain: Antiferromagnetism in K-Doped p-Terphenyl. *J. Am. Chem. Soc.* **2020**, *142*, 20624–20630. [CrossRef]
77. Márkus, B.G.; Szirmai, P.; Edelthalhammer, K.F.; Eckerlein, P.; Hirsch, A.; Hauke, F.; Nemes, N.M.; Chacón-Torres, J.C.; Náfrádi, B.; Forró, L.; et al. Ultralong Spin Lifetime in Light Alkali Atom Doped Graphene. *ACS Nano* **2020**, *14*, 7492–7501. [CrossRef]
78. Wang, R.-S.; Peng, D.; Zong, L.-N.; Zhu, Z.-W.; Chen, X.-J. Full set of superconducting parameters of K 3 C60. *Carbon N. Y.* **2023**, *202*, 325–335. [CrossRef]
79. Kytin, V.G.; Bulychev, B.M.; Krechetov, A.V.; Konstantinova, E.A.; Kulbachinskii, V.A.; Lunin, R.A. Investigation of paramagnetic centers in fullerides A 2MC 60 and AM 2C 60 (A = K, Rb, M = Mg, Be). *Appl. Magn. Reson.* **2008**, *33*, 177–184. [CrossRef]
80. Galambos, M.; Fábián, G.; Simon, F.; Ćirić, L.; Forró, L.; Korecz, L.; Rockenbauer, A.; Koltai, J.; Zólyomi, V.; Rusznyák, Á.; et al. Identifying the electron spin resonance of conduction electrons in alkali doped SWCNTs. *Phys. Status Solidi* **2009**, *246*, 2760–2763. [CrossRef]
81. Dediu, V.; Murgia, M.; Matacotta, F.C.; Taliani, C.; Barbanera, S. Room temperature spin polarized injection in organic semiconductor. *Solid State Commun.* **2002**, *122*, 181–184. [CrossRef]
82. Wittmann, A.; Schweicher, G.; Broch, K.; Novak, J.; Lami, V.; Cornil, D.; McNellis, E.R.; Zadvorna, O.; Venkateshvaran, D.; Takimiya, K.; et al. Tuning Spin Current Injection at Ferromagnet-Nonmagnet Interfaces by Molecular Design. *Phys. Rev. Lett.* **2020**, *124*, 027204. [CrossRef] [PubMed]
83. Yuan, J.; Zhang, Y.; Zhou, L.; Zhang, G.; Yip, H.L.; Lau, T.K.; Lu, X.; Zhu, C.; Peng, H.; Johnson, P.A.; et al. Single-Junction Organic Solar Cell with over 15% Efficiency Using Fused-Ring Acceptor with Electron-Deficient Core. *Joule* **2019**, *3*, 1140–1151. [CrossRef]

nanomaterials

Article

Mn-X (X = F, Cl, Br, I) Co-Doped GeSe Monolayers: Stabilities and Electronic, Spintronic and Optical Properties

Wenjie He, Xi Zhang *, Dan Gong, Ya Nie and Gang Xiang *

College of Physics, Sichuan University, Chengdu 610065, China; 13980505233@163.com (W.H.); 15183568892@163.com (D.G.); nieya1104@scu.edu.cn (Y.N.)
* Correspondence: xizhang@scu.edu.cn (X.Z.); gxiang@scu.edu.cn (G.X.)

Abstract: GeSe monolayer (ML) has recently attracted much interest due to its unique structure and excellent physical properties that can be effectively tuned through single doping of various elements. However, the co-doping effects on GeSe ML are rarely studied. In this study, the structures and physical properties of Mn-X (X = F, Cl, Br, I) co-doped GeSe MLs are investigated by using first-principle calculations. The formation energy and phonon disspersion analyses reveal the stability of Mn-Cl and Mn-Br co-doped GeSe MLs and instability of Mn-F and Mn-I co-doped GeSe MLs. The stable Mn-X (X = Cl, Br) co-doped GeSe MLs exhibit complex bonding structures with respect to Mn-doped GeSe ML. More importantly, Mn-Cl and Mn-Br co-doping can not only tune magnetic properties, but also change the electronic properties of GeSe MLs, which makes Mn-X co-doped GeSe MLs indirect band semiconductors with anisotropic large carrier mobility and asymmetric spin-dependent band structures. Furthermore, Mn-X (X = Cl, Br) co-doped GeSe MLs show weakened in-plane optical absorption and reflection in the visible band. Our results may be useful for electronic, spintronic and optical applications based on Mn-X co-doped GeSe MLs.

Keywords: GeSe; co-doping; carrier mobility; magnetization; optical property

check for updates

Citation: He, W.; Zhang, X.; Gong, D.; Nie, Y.; Xiang, G. Mn-X (X = F, Cl, Br, I) Co-Doped GeSe Monolayers: Stabilities and Electronic, Spintronic and Optical Properties. *Nanomaterials* **2023**, *13*, 1862. https://doi.org/10.3390/nano13121862

Academic Editors: Cesare Malagù and Antonino Gulino

Received: 11 May 2023
Revised: 10 June 2023
Accepted: 12 June 2023
Published: 15 June 2023

1. Introduction

Since the experimental realization of graphene [1], two-dimensional (2D) materials have attracted much interest and have been heavily investigated, such as black phosphorene (BP) [2], hexagonal boron nitride (h-BN) [3], group IV monolayers (MLs) [4], transition-metal dichalcogenide (TMD) MLs [4], metal dichalcogenide (MD) MLs [5,6], group IV-V MLs [7], group IV monochalcogenide MLs [8], group III-V MLs [9] and ternary Bi oxyhalide MLs [10]. Among them, with a similar structure to that of BP, group VI monochalcogenide (MX, M = Ge, Sn; X = S, Se) MLs have attracted increasing attention for their excellent optical and electrical properties [8,11–14]. In contrast to most MX MLs with indirect band gaps, GeSe ML is a p-type semiconductor with a direct band gap of 1.16 eV [12,15] and exhibits a carrier mobility as high as 6.22×10^3 cm^{2}v^{-1}s^{-1} [11], which is desirable for electrical and optical devices. Furthermore, the physical properties of GeSe ML can be modulated effectively through single doping of diverse elements. For instance, GeSe ML can be turned from nonmagnetic to magnetic through single doping of transition-metal (TM) elements [16], among which Mn doping introduces the largest magnetic moment of 5.0 μB per dopant [17]. Meanwhile, single doping of halogen element X (X = F, Cl, Br, I) can significantly modify the electrical properties of GeSe ML by introducing additional carriers [18]. It is noted that Li et al. studied the co-doping effect of Zn-Ga on the thermoelectric properties of 2D group VI monochalcogenide SnSe nanosheets [19]. However, there are no reports on Mn-X co-doping effects on the electronic, magnetic and optical properties of 2D GeSe MLs, which may be useful for the design and application of GeSe-based materials and devices.

In this work, we have investigated the stabilities and structures and electronic, magnetic and optical properties of Mn-X (X = F, Cl, Br, I) co-doped GeSe MLs by using first-principle calculations. The phonon dispersion curves revealed the stability of Mn-Cl and Mn-Br co-doped GeSe MLs and the instability of Mn-F and Mn-I co-doped GeSe MLs. Further calculations showed that the magnetic, electronic and optical properties of GeSe MLs were tuned by the Mn-X co-doping simultaneously. Our results provide insight into the Mn-X co-doping effects on the GeSe MLs and may be useful for the design, fabrication and application of Mn-X co-doping group IV monochalcogenide MLs.

2. Computational Methods

The calculations were performed using the Vienna ab initio Simulation Package (VASP) based on density functional theory (DFT) [20]. Both the properties of undoped and doped GeSe MLs were optimized with the projected augmented wave (PAW) method [21]. The Perdew-Burke-Ernzerhof (PBE) function of the generalized gradient approximation (GGA) was used as the exchange correlation function [22]. The simulations were done in a 2×2 supercell GeSe ML with 8 Ge and 8 Se atoms, in which 1 Mn and 1 X atoms are doped, corresponding to a doping concentration of 6.25% for each element. We chose a 3×3 supercell in addition to a 2×2 supercell to perform the calculations to evaluate the influence of material size and doping concentration. We used the projector augmented-wave pseudopotential method with a plane-wave basis set with a kinetic-energy cutoff of 500 eV. The Monkhorst-Pack scheme was used in the Brillouin zone with k-point meshes of $11 \times 11 \times 1$ throughout the calculations. The thickness of the vacuum region was kept greater than 15 Å. Structural relaxation was carried out until the total energy converged to 10^{-6} eV and the force on each atom was less than 0.01 eV/Å. Gaussian smearing was used with a small width of 0.01 eV. For describing the strong correlation interaction of d electrons, the GGA + U method was used with a U value of 3.0 eV for the Mn atom [16].

3. Results and Discussions

Figure 1a shows the structure of a 2×2 GeSe ML supercell. The lattice constants were calculated to be a = 3.99 Å and b = 4.26 Å in GeSe primitive cell, as shown in Table 1, consistent with previous theoretical results [12,23], although slightly different from experimental results [24], in which b (10.840 Å) is the lattice parameter along the layers, while a (4.394 Å) corresponds to our b (4.26 Å) and c (3.833 Å) corresponds to our a (3.99 Å). The Mn-X co-doping process was performed as follows. The Mn doping position was first optimized, and then the X doping position was considered and optimized, as shown in Figure 1b,c. The formation energy E_{form} of the co-doped GeSe ML was calculated using the following formula: [25] $E_{form} = E_{doped} - E_{undoped} + E_{sites} - E_{dopants}$, where E_{doped} and $E_{undoped}$ are the total energies of the GeSe ML with and without dopant atoms, respectively, E_{sites} is the sum of the energy of substituted Ge and Se atoms, and $E_{dopants}$ is the sum of the energy of the Mn and X dopants. It was found that the E_{form} value (0.85 eV) of Mn-doped GeSe ML obtained from Mn substituting Ge is smaller than that (3.33 eV) from Mn substituting Se, meaning that Mn tends to occupy the Ge site, consistent with previous theoretical studies [16]. Figure 1d shows that, given that Mn is substituting Ge, the formation energies of X atoms doped at the Se sites are generally lower than those at the Ge sites in the Mn-X co-doping, indicating that X atoms tend to occupy the Se sites. Specifically, F, Cl and Br tend to occupy the 1(Se) site (Figure 1c) adjacent to the Mn sites, while I tends to occupy the top (Se) site (Figure 1c) opposite to the Mn site, probably due to its larger atomic mass. Based on the results above, for convenience, Mn-doped GeSe ML is named (Ge, Mn) Se ML, while Mn-X co-doped GeSe ML is named (Ge, Mn) (Se, X) ML.

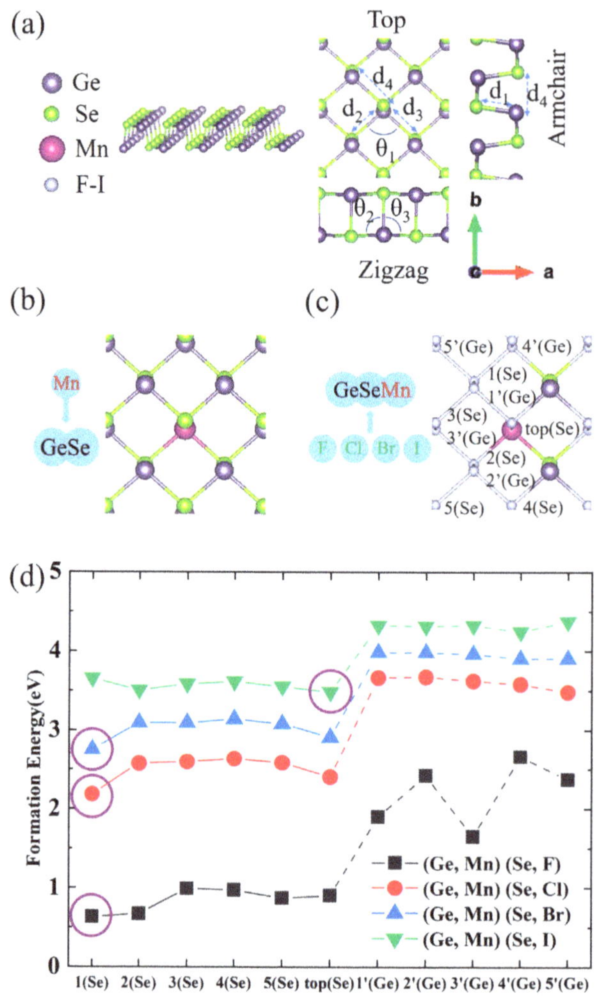

Figure 1. (**a**) The atomic structure of GeSe ML. (**b**) The doping site of Mn in Mn-doped GeSe ML. (**c**) The possible doping sites of X in Mn-X co-doped GeSe MLs. (**d**) The formation energies of Mn-X co-doped GeSe MLs as a function of doping sites. The doping sites with the lowest energy formation are circled in purple.

Table 1. The lattice constants (a and b), bond angles (θ_1, θ_2 and θ_3) and bond lengths (d_1, d_2, d_3 and d_4) of undoped and doped GeSe MLs.

Dopant	a (Å)	b (Å)	θ_1°	θ_2°	θ_3°	d_1 (Å)	d_2 (Å)	d_3 (Å)	d_4 (Å)
None	7.97	8.52	96.38	97.53	97.53	2.55	2.68	2.68	
Mn	7.96	8.27	92.96	101.55	101.55	2.57	2.68	2.68	2.86
Mn-Cl	7.84	9.19	89.60	108.49	97.18	2.59	2.62	2.79	2.61
Mn-Br	7.87	9.17	90.42	107.85	97.70	2.59	2.61	2.78	2.79

To further explore the thermodynamic stability of undoped, Mn-doped and Mn-X co-doped GeSe MLs, the phonon dispersion curves were calculated by using VASP and PHONOPY packages based on the density functional perturbation theory (DFPT), ref. [26]

as shown in Figure 2a–f, where the absence of imaginary frequencies indicates the stability of the materials. The phonon dispersion curves of GeSe ML show three acoustic branches and nine optical branches, which is consistent with previous theoretical calculations [27,28]. As shown in Figure 2, all undoped and doped GeSe MLs are stable, except for Mn-F and Mn-I co-doped GeSe MLs. Their instability may be attributed to the overly active outer electrons of the F atom and the large effective mass of the I atom. From the phonon spectra, it was observed that Mn-F and Mn-I co-doping rendered GeSe monolayers unstable. This observation may save experimenters time in related attempts and also remind them to choose doping elements more carefully in similar co-doping systems. Therefore, the relevant properties of Mn-F and Mn-I co-doped GeSe MLs will not be discussed.

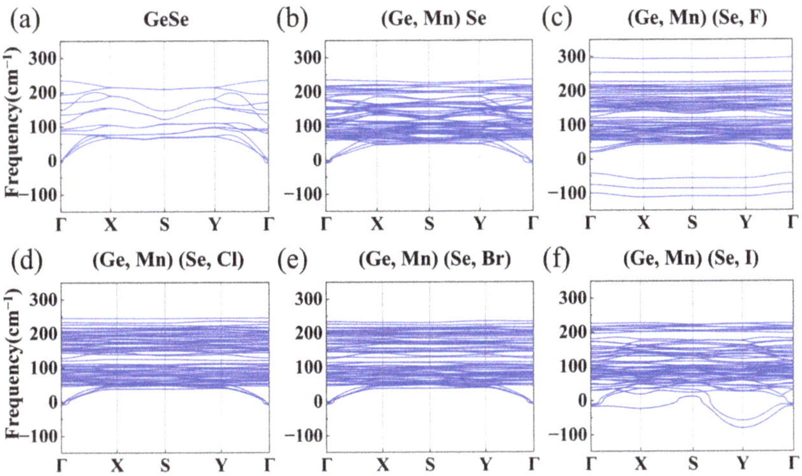

Figure 2. (**a**–**f**) The phonon dispersion curves of undoped, Mn-doped and Mn-X co-doped GeSe MLs.

Subsequently, the doping effects on the structure of GeSe MLs were investigated. The calculated structural parameters of undoped and doped GeSe MLs are listed in Table 1. The changes in lattice constants a and b indicated that Mn-X co-doping generally had a larger impact on the structure than single doping of Mn, which is also shown in Figure 2b–f. In addition, (Ge, Mn) Se ML's d_2 and d_3 were equal to GeSe ML's d_2 and d_3, while Mn-X co-doped GeSe MLs showed obvious variations in d_2 and d_3, again suggesting that co-doping has a stronger effect on the structure of GeSe ML. d_4, the bond length between Mn atom and X atom in (Ge, Mn) (Se, X) (X = Cl, Br) ML, was 2.61 Å and 2.79 Å for Mn-Cl and Mn-Br co-doping, respectively, revealing the influence of electronegativity of halogen elements on the structure of GeSe MLs.

The bonding structures of undoped and doped GeSe MLs were revealed by charge density differences ($\Delta\rho = \rho_{doped} - \rho_{vacancy} - \rho_{dopant}$), as shown in Figure 3a–d. In single Mn doping, the Mn atom formed five bonds with the surrounding Ge atoms. In Mn-X co-doping, the Mn atom formed five bonds with an X atom and four Se atoms, while the X atom formed two bonds with the Mn and Ge atoms. When the X atom replaced a Se atom, the Ge-Se bonds were broken, and the Mn-X bond was formed, which further rearranged the charge density and affected the structure. The structure of the Mn-X co-doped GeSe monolayer was more variable than that of Mn single doping. The position of the X atom changed to a certain extent, and a hole formed around it due to the lack of bonds, which resulted in more peculiar properties due to structural changes. In addition, some new bonds were formed far away from the Mn and X (Cl, Br) atoms, which may explain (Ge, Mn) (Se, X) ML's strange electrical and magnetic properties.

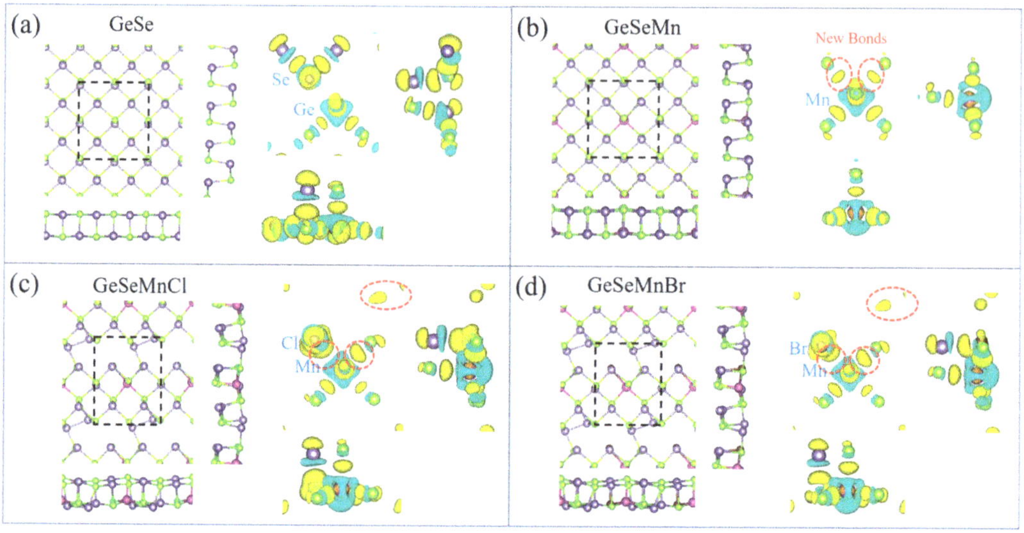

Figure 3. (**a–d**) The fully optimized structures and charge density differences of undoped, Mn-doped, Mn-Cl co-doped and Mn-Br co-doped GeSe MLs, respectively. The supercells are outlined using a black-dotted box. The new bands are outlined using red-dotted circles. Navy balls, green balls and purple balls denote Ge atoms, Se atoms and Mn atoms, respectively, while the other colors represent different X atoms. For charge density differences, the isosurface value is selected at 0.004 eV/Å. The yellow color and blue color denote electron the accumulation region and depletion region, respectively.

The electronic band structures of the undoped, Mn-doped and Mn-X (Cl, Br) co-doped GeSe MLs were then studied. Figure 4a shows that GeSe ML possesses a direct band gap of 1.13 eV, where the valence-band maximum (VBM) and the conduction-band minimum (CBM) lie along the Γ-Y direction, consistent with previous theoretical results (1.16 eV) [12,15]. The spin dependent band structures in Figure 4b–d show that (Ge, Mn) Se, (Ge, Mn) (Se, Cl) and (Ge, Mn) (Se, Br) MLs are indirect semiconductors. As shown in Figure 4 and Table 2, the up-spin bands of (Ge, Mn) Se, (Ge, Mn) (Se, Cl) and (Ge, Mn) (Se, Br) MLs exhibit band gap values of 0.95 eV, 0.58 eV and 0.57 eV, respectively, smaller than the band gap value of GeSe ML, while both (Ge, Mn) (Se, Cl) and (Ge, Mn) (Se, Br) MLs are tuned to be n-type semiconductors. The down-spin bands of (Ge, Mn) Se, (Ge, Mn) (Se, Cl) and (Ge, Mn) (Se, Br) MLs exhibit band gaps of 1.08 eV, 1.07 eV and 1.08 eV, respectively. The asymmetry between the up-spin and down-spin band structures of (Ge, Mn) (Se, Cl) and (Ge, Mn) (Se, Br) MLs demonstrates their potential in semiconductor spintronic applications.

To investigate the effect of different U values on the electronic structure of our systems, the bands of Mn-doped, Mn-Cl and Mn-Br co-doped GeSe MLs with the previously used U values of 4.0 eV, 5.0 eV and 6.0 eV [29–31], in addition to the U value of 3.0 eV, were calculated and shown in Figure S1. The band structures were nearly consistent, indicating that our systems were insensitive to the U values above 3.0 eV. For a clearer comparison, the bandgap values and their relative change ratios are listed in Table S1, where the maximum change ratio is as low as 2.88%, indicating that a U value of 3.0 eV is sufficient for the Mn atom in our systems.

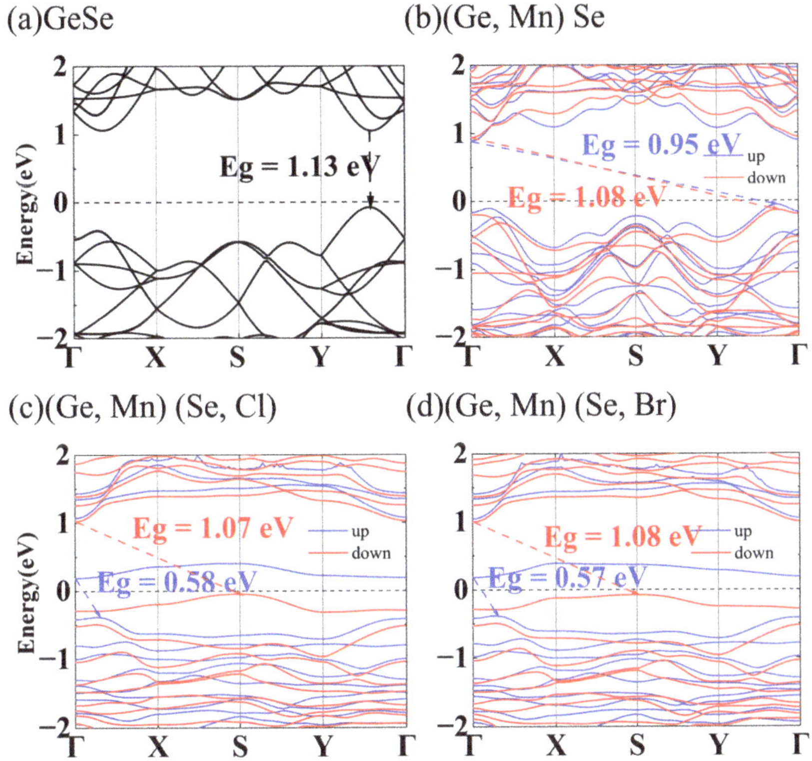

Figure 4. The band structures of (**a**) undoped, (**b**) Mn-doped, (**c**) Mn-Cl co-doped and (**d**) Mn-Br co-doped GeSe MLs.

Table 2. The band gaps of undoped and doped GeSe MLs, where D and I demonstrate direct and indirect band gaps, respectively.

Dopant	Spin	Bandgap (eV)
None	up	1.13(D)
	down	1.13(D)
Mn	up	0.95(I)
	down	1.08(I)
Mn-Cl	up	0.58(I)
	down	1.07(I)
Mn-Br	up	0.57(I)
	down	1.08(I)

In order to clarify contributions from different orbitals to band structures around the Fermi level, the partial density of states (PDOSs) of undoped and doped GeSe MLs are shown in Figure 5. In GeSe ML, CBM and VBM were contributed by Ge p and Se p orbital, respectively. In (Ge, Mn) Se ML, CBM and VBM of the up-spin bands were contributed by Mn s and d orbitals, respectively, while CBM and VBM of the down-spin bands were contributed by Mn d orbital. Ge and Se p orbitals hybridized with the Mn d orbital and contributed significantly to the CBM and VBM. In (Ge, Mn) (Se, Cl) and (Ge, Mn) (Se, Br) MLs, impurity levels were contributed by Cl p and Br p orbitals, and were close to the CBMs in the up-spin bands and close to the VBMs in the down-spin bands, respectively.

Ge and Se p orbitals also played an important role in impurity levels due to hybridization with Ge, Se, X (Cl, Br) p orbitals and Mn d orbital.

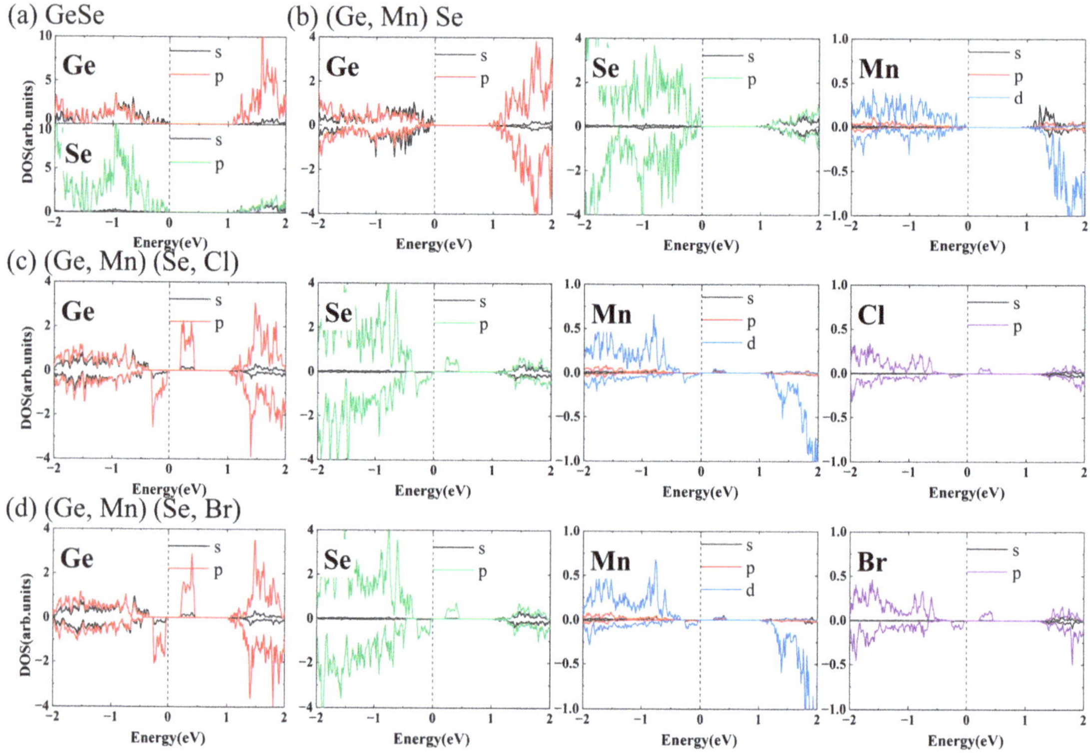

Figure 5. PDOSs of (**a**) undoped, (**b**) Mn-doped, (**c**) Mn-Cl co-doped and (**d**) Mn-Br co-doped GeSe MLs.

Since undoped, Mn-doped, Mn-Cl co-doped and Mn-Br co-doped GeSe MLs are semiconductors and carrier mobility is one of the key parameters of semiconductors [32], the electron and hole carrier mobility values of them were calculated using the deformation potential theory two dimensional (2D) semiconductors [33,34], as follows:

$$\mu_{2D} = \frac{2e\hbar^3 C_{2D}}{3k_B T m_e^* m_d E_l^2}$$

where C_{2D} is elastic modulus of the strain, derived from $C_{2D} = \frac{\left(\frac{\partial^2 E}{\partial^2 \left(\frac{\Delta l}{l_0}\right)}\right)}{S}$, in which E is total energy of systems; Δl is a small change in the lattice constant l_0; S is the area of lattice for a 2D system; T is temperature (300 K); m_e^* is the effective mass of the carrier along the transport direction; and m_d is the average effective mass, derived from $m_d = \sqrt{(m_a^* m_b^*)}$; and E_l is the deformation potential, derived from $E_l = \frac{\Delta E}{\left(\frac{\Delta l}{l_0}\right)}$, in which ΔE is energy change of CBM or VBM under strain. As shown in Table 3, the carrier mobility of GeSe ML in the b direction reaches 7494 $cm^2v^{-1}s^{-1}$, similar to previous theoretical results [11,12]. The maximum carrier mobility values of (Ge, Mn) Se, (Ge, Mn) (Se, Cl) and (Ge, Mn) (Se, Br) MLs were 5603 $cm^2v^{-1}s^{-1}$, 9060 $cm^2v^{-1}s^{-1}$, and 3652 $cm^2v^{-1}s^{-1}$, respectively, indicating that the MLs can keep high carrier mobility even after Mn-Cl and Mn-Br co-doping. The main carriers (holes) mobility values of (Ge, Mn) (Se, Cl) and (Ge, Mn) (Se, Br) MLs in a spin-down state were 107 $cm^2v^{-1}s^{-1}$, 14 $cm^2v^{-1}s^{-1}$ in the a direction and 16 $cm^2v^{-1}s^{-1}$, 2 $cm^2v^{-1}s^{-1}$ in the b direction, respectively, The main carrier (electrons) mobility values of (Ge, Mn) (Se, Cl) and (Ge, Mn) (Se, Br) MLs in a spin-up state were 673 $cm^2v^{-1}s^{-1}$, 2987 $cm^2v^{-1}s^{-1}$ in the a direction and and 8 $cm^2v^{-1}s^{-1}$, 20 $cm^2v^{-1}s^{-1}$ in the b direction, respectively, indicating the anisotropic nature of carrier mobility following Mn-Cl and Mn-Br co-doping. Interestingly, following Mn-Br co-doping, GeSe ML's main carriers in a spin-up state changed from holes (450 $cm^2v^{-1}s^{-1}$) to electrons (2987 $cm^2v^{-1}s^{-1}$).

Table 3. Effective mass m (with m_0 being the static electron mass), deformation potential constant E, elastic modulus C, and carrier mobility μ along the a and b directions. The electron and hole carrier mobility μ values are calculated at T = 300 K.

Dopant	Spin	Carrier	m_a (m_0)	m_b (m_0)	C_a (N/m)	C_b (N/m)	E_a (eV)	E_b (eV)	μ_a ($cm^2v^{-1}s^{-1}$)	μ_b ($cm^2v^{-1}s^{-1}$)
None		electron	0.37	0.13	45.56	22.65	6.77	1.22	172	7494
		hole	0.31	0.14	45.56	22.65	10.18	4.92	96	450
Mn	up	electron	0.14	0.18	49.84	35.13	4.26	2.02	1735	4230
		hole	0.32	0.18	49.84	35.13	7.06	5.80	183	339
	down	electron	0.16	0.19	49.84	35.13	2.59	1.63	3739	5603
		hole	0.30	0.24	49.84	35.13	7.66	4.76	148	338
Mn-Cl	up	electron	1.79	6.94	29.05	33.79	0.31	1.52	673	8
		hole	0.42	0.63	29.05	33.79	4.12	3.35	111	130
	down	electron	0.24	0.4	29.05	33.79	3.23	0.65	526	9060
		hole	1.39	2.27	29.05	33.79	1.24	2.69	107	16
Mn-Br	up	electron	1.57	1.88	30.36	17.58	0.23	1.96	2987	20
		hole	0.42	0.62	30.36	17.58	4.51	3.44	98	66
	down	electron	0.26	0.35	30.36	17.58	2.4	0.8	944	3652
		hole	2.58	6.39	30.36	17.58	1.68	2.42	14	2

The magnetic properties of the Mn-doped and Mn-X co-doped GeSe MLs were then investigated. The calculated magnetic moments and spin charge density ($\Delta\rho = \rho_{up} - \rho_{down}$) distributions in Figure 6 show that all the doped GeSe MLs are magnetic. Our results indicate that single doping of Mn results in a magnetic moment of 5.0 μB in (Ge, Mn) Se ML, which mainly comes from the spin-up electrons around Mn, consistent with previous theoretical results [16,17] and slightly larger than the 4.25 μB observed in GeMnSe nanocombs in previous experimental studies [35]. Mn-Cl and Mn-Br co-doping further weakened the magnetic moment to 4.0 μB in both cases by introducing more spin-down electrons in (Ge, Mn) (Se, Cl) and (Ge, Mn) (Se, Br) MLs. The spin-down electrons that produce changes at sites far away from the Mn and X (Cl, Br) atoms significantly correlated with new bonds, as shown in Figure 3.

The optical properties of undoped and doped GeSe MLs were then investigated. The optical absorption and reflectivity of undoped and doped GeSe MLs were almost the same in the a and b directions (the in-plane directions), as shown in Figure 7. The optical properties of Mn-doped GeSe ML were similar to those of GeSe ML. With respect to those of undoped GeSe ML, the in-plane optical absorption and reflection of Mn-Cl and Mn-Br co-doped GeSe MLs in the visible band were weakened but remained at the same order of magnitude. The optical absorption and reflection of Mn-Br co-doped GeSe ML were slightly stronger than those of Mn-Cl co-doped GeSe ML. Weaker optical absorption and reflection in the visible band suggest that Mn-X co-doped GeSe MLs could be useful as transparent materials. In the ultraviolet band, the in-plane optical absorption of Mn-Cl and Mn-Br co-doped GeSe MLs was strengthened, while the in-plane optical reflection remained weak. Stronger optical absorption and weaker reflection in the ultraviolet band suggested that Mn-X co-doped GeSe MLs could be useful for ultraviolet light absorption.

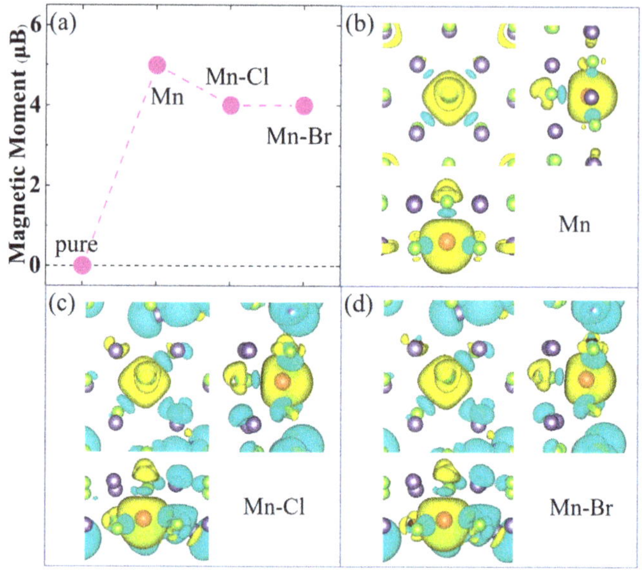

Figure 6. (**a**) Magnetic moments as the function of the undoped and doped GeSe MLs. Spin charge density of (**b**) Mn-doped, (**c**) Mn-Cl co-doped and (**d**) Mn-Br co-doped GeSe MLs. The yellow and blue represent the spin-up and spin-down electrons distribution, and the isosurface value is set at 0.001 eV/Å.

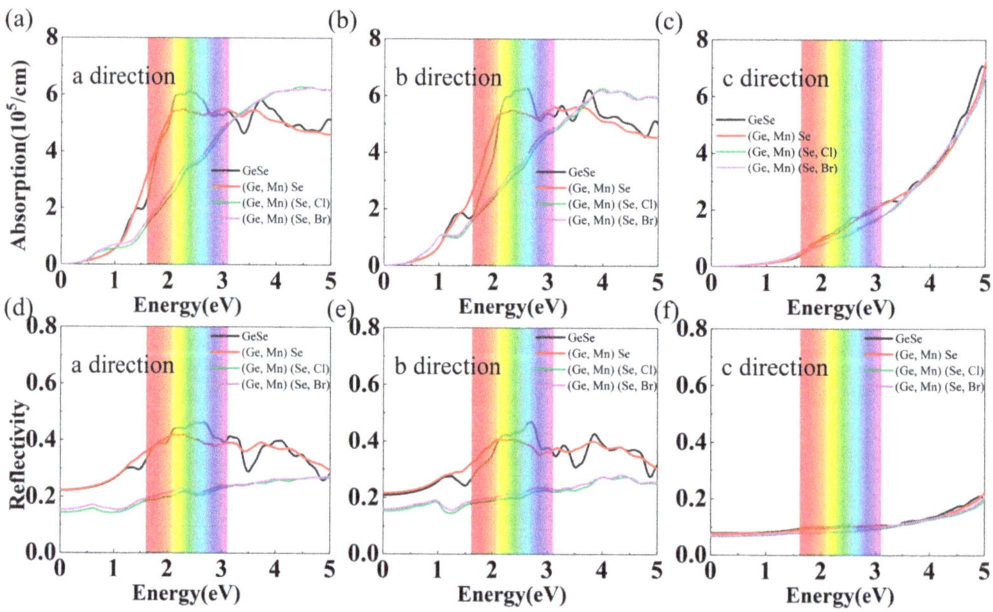

Figure 7. Light absorption of undoped, Mn-doped and Mn-X co-doped GeSe MLs along (**a**) the a direction, (**b**) the b direction and (**c**) the c direction. Light reflectivity of undoped, Mn-doped and Mn-X co-doped GeSe MLs along (**d**) the a direction, (**e**) the b direction and (**f**) the c direction.

Finally, to evaluate the influence of material size and doping concentration, we chose a 3×3 supercell in addition to a 2×2 supercell to perform the calculations. We calculated undoped, Mn-doped, and Mn-X co-doped GeSe monolayers in 3×3 supercells, which corresponded to a doping concentration of 2.78% for each element. We did not change the parameters of the calculation, except for using the Monkhorst-Pack scheme in the Brillouin zone with k-point meshes of $7 \times 7 \times 1$ throughout all the calculations in 3×3 supercells, to ensure consistent accuracy. Then the band structures of the undoped and doped GeSe monolayers under the 2×2 and 3×3 supercells were compared, as shown in Figure S2. For GeSe ML in 3×3 supercells, compared to that in 2×2 supercells, the bandgap type remained direct and the bandgap changed slightly from 1.13 eV to 1.14 eV. For doped GeSe monolayers, the bandgap type remained indirect. However, for Mn-doped, Mn-Cl co-doped and Mn-Br co-doped GeSe Monolayers, the bandgap changed by 0.15 eV, 0.08 eV and 0.09 eV in the spin-up states, and by 0.12 eV, 0.09 eV and 0.12 eV in spin-down states, respectively. After enlarging the supercells, the variations of the band gaps ranged from 8.4% to 15.8%, indicating that the doping concentration had a small effect on the bandgap, as shown in Table S2. To examine the contributions of different orbitals to the band structures, the PDOS of undoped and doped GeSe Monolayers were studied using supercells of two sizes, as shown in Figure S3. We found that since the numbers of Ge and Se atoms were almost doubled, the contributions of their orbitals to the bands were also nearly doubled. From the PDOS, our conclusions about which elements and which orbitals contributed to the VBM and CBM remain unchanged. It is worth noting that the contributions of the Ge p and Se p orbitals to the impurity levels were not doubled, indicating that the asymmetric spin-dependent band structures were mainly derived from the Mn and X atoms and their nearby Ge and Se atoms. The change in doping concentration did not change this characteristic feature. The spin charge densities of undoped and doped GeSe Monolayers are shown in Figure S4. In 3×3 supercells, the magnetic moment of Mn-doped GeSe ML mainly came from the spin-up electrons around Mn. Mn-Cl and Mn-Br co-doping introduced more spin-down electrons at sites far away from the Mn and X (Cl, Br) atoms. In different supercells, our conclusions regarding the origin of the magnetic properties in doped GeSe monolayers remain unchanged. As shown in Table S3, the magnetic moments remain consistent in different sized supercells, which are 5 μB, 4 μB and 4 μB for Mn doped, Mn-Cl co-doped and Mn-Br co-doped GeSe monolayers, respectively. Finally, we investigated the light absorption and reflection of undoped and doped GeSe monolayers in different sized supercells, as shown in Figure S5. In 3×3 supercells, the reduced light absorption and reflection capacity in the visible band of Mn-X co-doped GeSe monolayers were restored, meaning that the optical properties of Mn-X co-doped GeSe monolayers were sensitive to doping concentrations. Moreover, the differences in the optical properties of Mn-Cl and Mn-Br co-doped GeSe monolayers became more pronounced, indicating that optical properties with different doping elements had different sensitivity to doping concentrations. By tuning different doping concentrations, different optical properties can be achieved, showing potential in optoelectronic device applications.

4. Conclusions

In summary, we have systematically investigated the stabilities, structures and electrical, transport, magnetic and optical properties of transition metal Mn and halogen elements (X = F, Cl, Br, I) co-doped GeSe MLs by using first-principle calculations. Our results reveal the instability of Mn-X (X = F, I) co-doped GeSe MLs and the stability of Mn-X (X = Cl, Br) co-doped GeSe MLs. Further calculations show that Mn-X (X = Cl, Br) co-doped GeSe monolayers are indirect band semiconductors with anisotropic large carrier mobility and asymmetric spin-dependent band structures. Compared to Mn doping, co-doping with Mn-Cl and Mn-Br can not only adjust the magnetic moments (from 5 μB to 4 μB) but also alter the electronic properties of GeSe monolayers. Furthermore, Mn-X (X = Cl, Br) co-doped GeSe monolayers exhibit weakened in-plane optical absorption and reflection in the visible band, which may be useful for optical applications. Our results give insights into the Mn-X co-doping effects on the structures and physical properties of GeSe MLs and may be useful for related electronic, spintronic and optical applications.

Supplementary Materials: The following supporting information can be downloaded at: https://www.mdpi.com/article/10.3390/nano13121862/s1. Figure S1: The band structures of (a) Mn doped, (b) Mn-Cl co-doped, and (c) Mn-Br co-doped GeSe Monolayers with different U values; Figure S2: In 2 × 2 and 3 × 3 supercell, the band structures of (a) undoped, (b) Mn-doped, (c) Mn-Cl co-doped and (d) Mn-Br co-doped GeSe Monolayers; Figure S3: In 2 × 2 and 3 × 3 supercell, PDOS of (a) undoped, (b) Mn-doped, (c) Mn-Cl co-doped and (d) Mn-Br co-doped GeSe Monolayers; Figure S4: Spin charge density of (a) Mn-doped, (b) Mn-Cl co-doped, (c) Mn-Br co-doped GeSe Monolayers in 2 × 2 supercell. Spin charge density of (d) Mn-doped, (e) Mn-Cl co-doped and (f) Mn-Br co-doped GeSe Monolayers in 3 × 3 supercell. The yellow and blue represent the spin-up and spin-down electrons distribution, and the isosurface value is set at 0.001 eV/Å; Figure S5: In 2 × 2 and 3 × 3 supercell, light absorption of undoped, Mn-doped and Mn-X co-doped GeSe Monolayers along (a) the a direction, (b) the b direction and (c) the c direction. Light reflectivity of undoped, Mn-doped and Mn-X co-doped GeSe Monolayers along (d) the a direction, (e) the b direction and (f) the c direction; Table S1: The band gaps of Mn doped, Mn-Cl, and Mn-Br co-doped GeSe Monolayers with different U values. I represents indirect band gap; Table S2: In 2 × 2 and 3 × 3 supercell, the band gaps of undoped and doped GeSe Monolayers, where D and I show direct and indirect band gaps; Table S3: The magnetic moments of doped GeSe Monolayers.

Author Contributions: Conceptualization, W.H. and X.Z.; methodology, W.H., X.Z. and G.X.; software, W.H., D.G. and Y.N.; validation, W.H., X.Z. and G.X.; formal analysis, W.H., X.Z. and G.X.; investigation, W.H., D.G. and Y.N.; resources, X.Z. and G.X.; data curation, W.H.; writing—original draft preparation, W.H., X.Z. and G.X.; writing—review and editing, W.H. and G.X.; visualization, W.H.; supervision, G.X.; project administration, G.X.; funding acquisition, G.X. All authors have read and agreed to the published version of the manuscript.

Funding: This work was supported by National Key Research and Development Program of China (MOST) (Grant No. 2022YFA1405100) and National Natural Science Foundation of China (NSFC) (Grant No. 52172272).

Data Availability Statement: The data supporting the findings of this study are available from the authors upon reasonable and appropriate request.

Conflicts of Interest: The authors declare no conflict of interest.

References

1. Novoselov, K.S.; Geim, A.K.; Morozov, S.V.; Jiang, D.; Zhang, Y.; Dubonos, S.V.; Grigorieva, I.V.; Firsov, A.A. Electric Field in Atomically Thin Carbon Films. *Science* **2004**, *306*, 666–669. [CrossRef]
2. Liu, H.; Neal, A.T.; Zhu, Z.; Luo, Z.; Xu, X.; Tománek, D.; Ye, P.D. Phosphorene: An Unexplored 2D Semiconductor with a High Hole Mobility. *ACS Nano* **2014**, *8*, 4033–4041. [CrossRef]
3. Song, L.; Ci, L.; Lu, H.; Sorokin, P.B.; Jin, C.; Ni, J.; Kvashnin, A.G.; Kvashnin, D.G.; Lou, J.; Yakobson, B.I.; et al. Large Scale Growth and Characterization of Atomic Hexagonal Boron Nitride Layers. *Nano Lett.* **2010**, *10*, 3209–3215. [CrossRef]
4. Mak, K.F.; Lee, C.; Hone, J.; Shan, J.; Heinz, T.F. Atomically Thin MoS2: A New Direct-Gap Semiconductor. *Phys. Rev. Lett.* **2010**, *105*, 136805. [CrossRef] [PubMed]
5. Ao, L.; Xiao, H.Y.; Xiang, X.; Li, S.; Liu, K.Z.; Huang, H.; Zu, X.T. Functionalization of a GaSe Monolayer by Vacancy and Chemical Element Doping. *Phys. Chem. Chem. Phys.* **2015**, *17*, 10737–10748. [CrossRef]
6. Ma, Y.; Dai, Y.; Guo, M.; Yu, L.; Huang, B. Tunable Electronic and Dielectric Behavior of GaS and GaSe Monolayers. *Phys. Chem. Chem. Phys.* **2013**, *15*, 7098–7105. [CrossRef] [PubMed]
7. Özdamar, B.; Özbal, G.; Çınar, M.N.; Sevim, K.; Kurt, G.; Kaya, B.; Sevinçli, H. Structural, Vibrational, and Electronic Properties of Single-Layer Hexagonal Crystals of Group IV and V Elements. *Phys. Rev. B* **2018**, *98*, 045431. [CrossRef]
8. Tyagi, K.; Waters, K.; Wang, G.; Gahtori, B.; Haranath, D.; Pandey, R. Thermoelectric Properties of SnSe Nanoribbons: A Theoretical Aspect. *Mater. Res. Express* **2016**, *3*, 035013. [CrossRef]
9. Gajić, R.; Meisels, R.; Kuchar, F.; Hingerl, K. All-Angle Left-Handed Negative Refraction in Kagomé and Honeycomb Lattice Photonic Crystals. *Phys. Rev. B Condens. Matter Mater. Phys.* **2006**, *73*, 165310. [CrossRef]
10. Gao, B.; Zhang, J.R.; Chen, L.; Guo, J.; Shen, S.; Au, C.T.; Yin, S.F.; Cai, M.Q. Density Functional Theory Calculation on Two-Dimensional MoS2/BiOX (X = Cl, Br, I) van Der Waals Heterostructures for Photocatalytic Action. *Appl. Surf. Sci.* **2019**, *492*, [illegible] [CrossRef]

11. Xu, L.; Yang, M.; Wang, S.J.; Feng, Y.P. Electronic and Optical Properties of the Monolayer Group-IV Monochalcogenides MX (M = Ge, Sn; X = S, Se, Te). *Phys. Rev. B* **2017**, *95*, 235434. [CrossRef]
12. Xu, Y.; Zhang, H.; Shao, H.; Ni, G.; Li, J.; Lu, H.; Zhang, R.; Peng, B.; Zhu, Y.; Zhu, H.; et al. First-Principles Study on the Electronic, Optical, and Transport Properties of Monolayer α- and β-GeSe. *Phys. Rev. B* **2017**, *96*, 245421. [CrossRef]
13. Yap, W.C.; Yang, Z.; Mehboudi, M.; Yan, J.A.; Barraza-Lopez, S.; Zhu, W. Layered Material GeSe and Vertical GeSe/MoS2 p-n Heterojunctions. *Nano Res.* **2018**, *11*, 420–430. [CrossRef]
14. Vaughn, D.D.; Patel, R.J.; Hickner, M.A.; Schaak, R.E. Single-Crystal Colloidal Nanosheets of GeS and GeSe. *J. Am. Chem. Soc.* **2010**, *132*, 15170–15172. [CrossRef]
15. Hu, Y.; Zhang, S.; Sun, S.; Xie, M.; Cai, B.; Zeng, H. GeSe Monolayer Semiconductor with Tunable Direct Band Gap and Small Carrier Effective Mass. *Appl. Phys. Lett.* **2015**, *107*, 122107. [CrossRef]
16. Zhao, Y.; Zhang, Y.; Yan, P.; Tu, J.; Xu, Y.; He, L. Increasing the Photocatalytic Properties of Monolayer Black Phase GeSe by 3d Transition Metal Doping: From Ultraviolet to Infrared Absorption. *Mol. Catal.* **2020**, *496*, 111195. [CrossRef]
17. Yang, L.; Wu, M.; Yao, K. Transition-Metal-Doped Group-IV Monochalcogenides: A Combination of Two-Dimensional Triferroics and Diluted Magnetic Semiconductors. *Nanotechnology* **2018**, *29*, 215703. [CrossRef] [PubMed]
18. Zhao, Y.; Li, X.; Li, H.; He, L. Surface Doping of Nonmetal Atoms Enhances Photocatalytic Performance of Monolayer GeSe for Degradation of Organic Pollution. *Chem. Phys. Lett.* **2021**, *785*, 139156. [CrossRef]
19. Li, S.; Hou, Y.; Li, D.; Zou, B.; Zhang, Q.; Cao, Y.; Tang, G. Realization of High Thermoelectric Performance in Solution-Synthesized Porous Zn and Ga Codoped SnSe Nanosheets. *J. Mater. Chem. A Mater.* **2022**, *10*, 12429–12437. [CrossRef]
20. Kresse, G.; Furthmüller, J. Efficient Iterative Schemes for Ab Initio Total-Energy Calculations Using a Plane-Wave Basis Set. *Phys. Rev. B Condens. Matter Mater. Phys.* **1996**, *54*, 11169–11186. [CrossRef]
21. Blöchl, P.E. Projector Augmented-Wave Method. *Phys. Rev. B* **1994**, *50*, 17953–17979. [CrossRef] [PubMed]
22. Perdew, J.P.; Burke, K.; Ernzerhof, M. Generalized Gradient Approximation Made Simple. *Phys. Rev. Lett.* **1996**, *77*, 3865–3868. [CrossRef]
23. Guo, Y.; Zhou, S.; Bai, Y.; Zhao, J. Oxidation Resistance of Monolayer Group-IV Monochalcogenides. *ACS Appl. Mater. Interfaces* **2017**, *9*, 12013–12020. [CrossRef]
24. von Rohr, F.O.; Ji, H.; Cevallos, F.A.; Gao, T.; Ong, N.P.; Cava, R.J. High-Pressure Synthesis and Characterization of β-GeSe—A Six-Membered-Ring Semiconductor in an Uncommon Boat Conformation. *J. Am. Chem. Soc.* **2017**, *139*, 2771–2777. [CrossRef] [PubMed]
25. Leenaerts, O.; Sahin, H.; Partoens, B.; Peeters, F.M. First-Principles Investigation of B- and N-Doped Fluorographene. *Phys. Rev. B Condens. Matter Mater. Phys.* **2013**, *88*, 035434. [CrossRef]
26. Baroni, S.; De Gironcoli, S.; Dal Corso, A.; Giannozzi, P. Phonons and Related Crystal Properties from Density-Functional Perturbation Theory. *Rev. Mod. Phys.* **2001**, *73*, 515–562. [CrossRef]
27. Chakraborty, R.; Ahmed, S.; Subrina, S. Functionalization of Electronic, Spin and Optical Properties of GeSe Monolayer by Substitutional Doping: A First-Principles Study. *Nanotechnology* **2021**, *32*, 305701. [CrossRef]
28. Shafique, A.; Shin, Y.H. Thermoelectric and Phonon Transport Properties of Two-Dimensional IV-VI Compounds. *Sci. Rep.* **2017**, *7*, 506. [CrossRef]
29. Dyachenko, A.A.; Lukoyanov, A.V.; Shorikov, A.O.; Anisimov, V.I. Magnetically Driven Phase Transitions with a Large Volume Collapse in MnSe under Pressure: A DFT+DMFT Study. *Phys. Rev. B* **2018**, *98*, 085139. [CrossRef]
30. Youn, S.J.; Min, B.I.; Freeman, A.J. Crossroads Electronic Structure of MnS, MnSe, and MnTe. *Phys. Status Solidi (b)* **2004**, *241*, 1411–1414. [CrossRef]
31. Hung, T.L.; Huang, C.H.; Deng, L.Z.; Ou, M.N.; Chen, Y.Y.; Wu, M.K.; Huyan, S.Y.; Chu, C.W.; Chen, P.J.; Lee, T.K. Pressure Induced Superconductivity in MnSe. *Nat. Commun.* **2021**, *12*, 5436. [CrossRef] [PubMed]
32. Park, S.Y.; Kim, P.J.; Lee, Y.P.; Shin, S.W.; Kim, T.H.; Kang, J.; Rhee, J.Y. Realization of Room-Temperature Ferromagnetism and of Improved Carrier Mobility in Mn-Doped ZnO Film by Oxygen Deficiency, Introduced by Hydrogen and Heat Treatments. *Adv. Mater.* **2007**, *19*, 3496–3500. [CrossRef]
33. Bardeen, J.; Shockley, W. Deformation Potentials and Mobilities in Non-Polar Crystals. *Phys. Rev.* **1950**, *80*, 72–80. [CrossRef]
34. Cai, Y.; Zhang, G.; Zhang, Y.W. Polarity-Reversed Robust Carrier Mobility in Monolayer MoS2 Nanoribbons. *J. Am. Chem. Soc.* **2014**, *136*, 6269–6275. [CrossRef]
35. Li, D.; Zhang, X.; He, W.; Lei, L.; Peng, Y.; Xiang, G. Structure-Dependent High-TC Ferromagnetism in Mn-Doped GeSe. *Nanoscale* **2022**, *14*, 13343–13351. [CrossRef] [PubMed]

nanomaterials

MDPI

Article

Controllable Synthesis and Charge Density Wave Phase Transitions of Two-Dimensional 1T-TaS_2 Crystals

Xiaoguang Pan, Tianwen Yang, Hangxin Bai, Jiangbo Peng, Lujie Li, Fangli Jing, Hailong Qiu, Hongjun Liu * and Zhanggui Hu

Tianjin Key Laboratory of Functional Crystal Materials, Institute of Functional Crystals, School of Materials Science and Engineering, Tianjin University of Technology, Tianjin 300384, China
* Correspondence: hjliu@email.tjut.edu.cn

Abstract: 1T-TaS_2 has attracted much attention recently due to its abundant charge density wave phases. In this work, high-quality two-dimensional 1T-TaS_2 crystals were successfully synthesized by a chemical vapor deposition method with controllable layer numbers, confirmed by the structural characterization. Based on the as-grown samples, their thickness-dependency nearly commensurate charge density wave/commensurate charge density wave phase transitions was revealed by the combination of the temperature-dependent resistance measurements and Raman spectra. The phase transition temperature increased with increasing thickness, but no apparent phase transition was found on the 2~3 nm thick crystals from temperature-dependent Raman spectra. The transition hysteresis loops due to temperature-dependent resistance changes of 1T-TaS_2 can be used for memory devices and oscillators, making 1T-TaS_2 a promising material for various electronic applications.

Keywords: 1T-TaS_2; CVD; resistance measurements; Raman spectra; NC/CCDW phase transition

Citation: Pan, X.; Yang, T.; Bai, H.; Peng, J.; Li, L.; Jing, F.; Qiu, H.; Liu, H.; Hu, Z. Controllable Synthesis and Charge Density Wave Phase Transitions of Two-Dimensional 1T-TaS_2 Crystals. *Nanomaterials* **2023**, *13*, 1806. https://doi.org/10.3390/nano13111806

Academic Editors: Anatoly Belonoshko and Seiichi Miyazaki

Received: 29 April 2023
Revised: 2 June 2023
Accepted: 2 June 2023
Published: 5 June 2023

1. Introduction

Two-dimensional (2D) layered materials have exhibited novel physical properties different from bulk materials due to their atomically low thickness and high carrier mobility. Among them, low-dimensional strongly correlated electron systems, such as 1T-TaS_2, 2H-$TaSe_2$, 2H-$NbSe_2$, and 1T-$TiSe_2$, have unique electronic structures and rich extraordinary physical properties, including superconductivity [1], charge density wave order [2,3], ferromagnetism [4], and catalytic activity [5]. As a charge density wave (CDW) material, 1T-TaS_2 has attracted much attention during the past years owing to its abundant phases, such as 1T, 2H, and 3R, with various stacking [6–10]. It has extensive prospects for applications as electronic, magnetic, and energy conversion devices, such as high-performance oscillators [6], fast memories [7], solar cells [8], humidity sensors [9,11–13], and high-efficiency electrocatalysts [10,14,15]. Recently, the ferroic character of 2D 1T-TaS_2 was established by revealing the hysteretic electrical switching of the ferro-rotational order through the observation of its domains and domain wall propagation [16].

Unlike the Peierls instability mechanism, electron–phonon interaction plays a crucial role in driving CDW instabilities [17,18]. The instability of CDW featured by periodic lattice distortion (PLD) was found to be dependent on the temperature, which affected spatial modulation of carrier density at low temperatures and produced a metastable phase [19–21]. Below 180 K, a commensurate CDW (CCDW) phase was revealed [22], in which a $\sqrt{13} \times \sqrt{13}$ superlattice was formed when 12 Ta atoms on the outside shrunk to the central 13th Ta atom to form a "Star of David" (SOD) structure [23]. After the CCDW phase transformation, the bandgap opening made 1T-TaS_2 a Mott insulator [24]. At 180 K, a nearly commensurate CDW (NCCDW) phase was formed with partial structures being commensurate to the original lattice, resulting in the metallic CCDW phase by reducing insulating domain walls. When further increasing to near 350 K, 1T-TaS_2 distorted into an incommensurate CDW (ICCDW) phase, in which atoms were shifted from their original lattice positions. In addition to thermal excitation, CDW phase transition can be induced in other ways, including photoinduced phase transition [25] and electron-induced phase transition [26,27]. Electrical measurement shows the sudden change in resistance and the hysteresis window [28].

In recent years, chemical vapor deposition (CVD) has been used as a mature strategy for controllable synthesis of high-quality 1T-TaS_2. The preparation of 2D TaS_2 crystals with different substrates has been reported, such as Au [29–31], SiO_2/Si [32], hexagonal boron nitride [33], sapphire [34], etc. However, the interaction between substrate and sample due to the charge transfer [35] inevitably affects CDW performance. Among these substrates, mica has excellent epitaxial growth characteristics and lattice adaptation degree. Here, we report the thickness-controllable growth of 1T-TaS_2 on mica substrates by ambient pressure CVD (APCVD) and studies on CDW phase transitions. The structure of 1T-TaS_2 was confirmed by Raman, X-ray diffraction (XRD), X-ray photoelectron spectroscopy (XPS), and high-resolution transmission electron microscope (HRTEM) measurements. After structural characterization, temperature-dependent resistance and temperature-varying Raman measurements were used to characterize the CDW phase transition under thermal excitation.

2. Materials and Methods

2.1. Characterizations of As-Grown 2D 1T-TaS_2 Crystals

Morphologies and thicknesses of 2D 1T-TaS_2 crystals were checked with an atomic force microscope (AFM, Bruker Corp., Billerica, MA, USA, Dimension Icon). The micro-Raman tests were performed with a confocal microscope-based Raman spectrometer (ALPHA 300, WITec Corp., Ulm, Germany) under an excitation laser at 532 nm. The temperature-dependent Raman spectra were collected in a custom-made vacuum thermostat ranging from 80 K to 260 K. The binding energies of elements were obtained by XPS measurements on as-grown samples (ESCALAB 250 Xi, Thermo Scientific Corp., Waltham, MA, USA). Before the TEM (Talos F200 X, FEI Corp., Hillsborough, OR, USA) measurements, the as-grown 1T-TaS_2 crystals were transferred onto micro-grid-supported Cu grids via a typical polymethyl methacrylate (PMMA)-assisted transfer method [36]. The as-grown samples were spin-coated with PMMA (950 K, A4, Allresist Corp., Straussberg, Berlin, Germany) at 6000 rpm for 60 s, followed by drying at 180 °C for 10 min. Then, the samples supported by PMMA film were lifted up with tweezers under deionized water, and then they were collected by micro-grid-supported Cu grids. Finally, the PMMA film was removed via dissolution with acetone for about 10 min and dried by flowing Ar gas. HRTEM images and selected area electron diffraction (SAED) patterns were collected by a Talos F200X transmission electron microscope operated at 200 kV.

2.2. Device Fabrication of 1T-TaS_2 Electrical Devices

For individual TaS_2 flakes, the electrodes were patterned by electron beam lithography (EBL). Five nanometers Ti and 50 nm Au were electron-beam evaporated for contacts. The devices were put in a vacuum (Janis ST500 probe station, $<10^{-5}$ Torr) and measured by an Agilent B1500A semiconductor device analyzer. Low-temperature resistance measurements were performed in a physical property measurement system (PPMS, Quantum Design, Inc., San Diego, CA, USA) under liquid He-purged conditions.

3. Results and Discussion

The top-view and side-view atomic models of 1T-TaS_2 are shown in Figure 1a,b. Ta atoms with the central octahedral arrangement are sandwiched between two S atom layers, demonstrating an ABC-type stacking [37]. The electron–phonon coupling-induced SOD structure at a low temperature is presented in Figure 1a, with red arrows indicating the shrinking direction of Ta atoms. At a temperature below 180 K, the 1T-TaS_2 crystal is filled with SOD cluster, while at room temperature, the 1T-TaS_2 crystal can transform from the CCDW to the NCCDW phase, which is only partially filled by satellite domains [38].

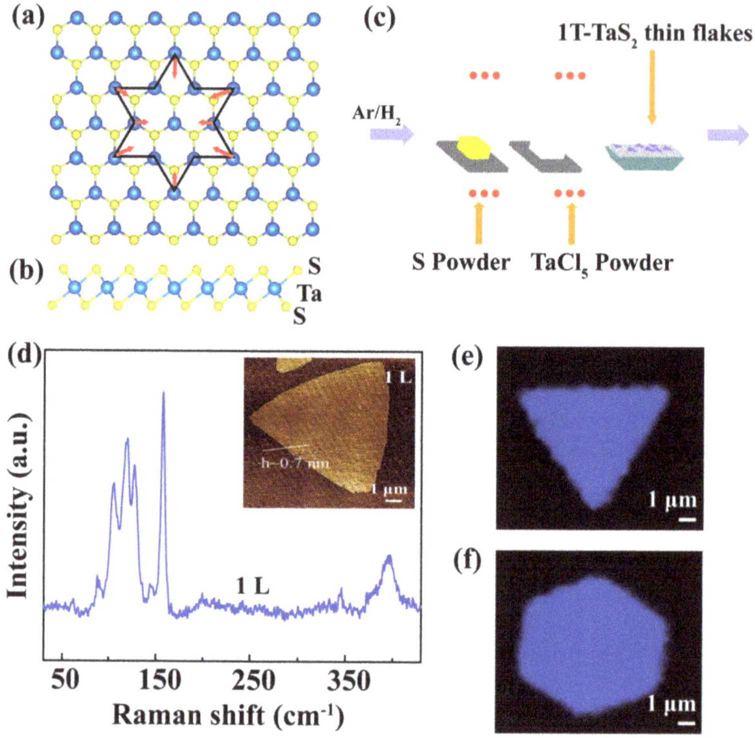

Figure 1. Synthesis of 1T-TaS_2 crystal via APCVD method. Top−view (**a**) and side view (**b**) of the atomic structures of a 1T-TaS_2 crystal. (**c**) Schematics for the APCVD growth of 1T-TaS_2 on a mica substrate. (**d**) A Raman spectrum collected at room temperature. An AFM image in the inset demonstrates the height of 0.7 nm for the monolayer, with a scale bar of 1 μm. (**e**,**f**) Raman maps of 1T-TaS_2 with different shapes.

Two-dimensional 1T-TaS_2 crystals were grown by the APCVD method, as schematically shown in Figure 1c. The mica was employed as a substrate placed in the deposition area. Tantalum pentachloride and sulfur powders were employed as the sources placed upstream outside the furnace, which were independently heated by two different heaters. Then, under the H_2/Ar (5%H_2) mixed carrier gas, the furnace was heated to 1100 K for growth. During the growth, the Ta flux was controlled while the S flux was kept continuous. By controlling the supply time for Ta and heating time, high-quality 1T-TaS_2 crystals with various layer thicknesses were grown. The Raman spectrum of 1T-TaS_2 monolayers collected at room temperature is shown in Figure 1d, in which peak wavenumbers less than 150 cm^{-1} were related to the tantalum atoms while phonon modes within 220−320 cm^{-1} were more associated with sulfur atoms [38]. The inset presents an AFM image of an as-grown 1T-TaS_2 crystal with a lateral size of about 10 μm on which the Raman spectrum was collected. The main Raman phonon modes of 1T-TaS_2 include A_{1g} modes of 71, 78, and 117 cm^{-1} and E_g modes of 60 and 90 cm^{-1}. Figure 1e,f shows Raman maps of 1T-TaS_2 crystals with unicolor trigonal and hexagonal shapes, indicating excellent crystallinity of the as-grown crystals.

The controllable synthesis of 1T-TaS_2 was further studied through AFM measurements, as shown in Figure 2a–d. The line profiles in the inset clearly show the thickness: from the monolayer of 0.7 nm to the tetralayer of 2.9 nm. It should be noted that the sizes of all grown 2D crystals were around 10 μm, which may be limited by the metallic precursor for this APCVD growth. With decreasing thickness, the stability of 1T-TaS_2 becomes worse.

The instability was further confirmed by the XPS measurements in Figure 3a. Besides the Ta-4f$_{7/2}$ peak at 24.2 eV for TaS_2, peaks of Ta-4f$_{5/2}$ at 26.8 eV, and peaks of Ta-4f$_{7/2}$ at 28.71 eV, a loss feature at about 38 eV for Ta_2O_5 also appears in the XPS spectrum, indicating the oxidization of 1T-TaS_2. The S-2p$_{1/2}$ and S-2p$_{3/2}$ are shown in Figure 3b, giving the evidence for TaS_2 [39]. For the monolayer, it is even difficult to perform Raman measurements on it due to the low laser-induced damage threshold. Thus, most of the measurements were performed on the thick 1T-TaS_2 crystals.

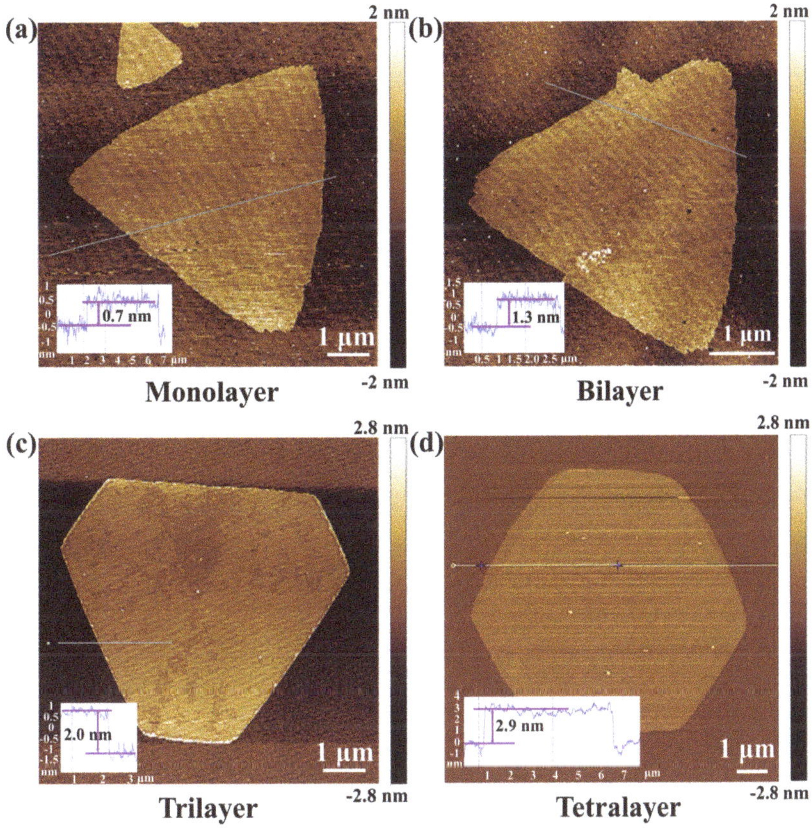

Figure 2. Layer—controlled synthesis of 1T-TaS_2 crystal via APCVD method. AFM images for the monolayer (**a**), bilayer (**b**), trilayer (**c**), and tetralayer (**d**) are presented, and the line profiles are shown in the corresponding inset, respectively. The scale bar is 1 μm in each image.

HRTEM measurements were performed to explore the internal structure and crystallinity of the sample. The as-grown 1T-TaS_2 was transferred to a micro-grid-supported carbon film via a PMMA-assisted transfer method [36]. As shown in Figure 3c, the low-magnification HRTEM image depicts the hexagonal shape of the sample with a uniform surface. HRTEM measurements were performed on the area marked in Figure 3c, and the corresponding atomic-resolution image is shown in Figure 3d. The SAED pattern shows the (100) and (001) with the spacing of 0.291 and 0.597 nm in the inset of Figure 3d, agreeing with the lattice spacing of 1T-TaS_2 [40]. In addition, the single clear dots indicate the high crystalline quality of the grown sample.

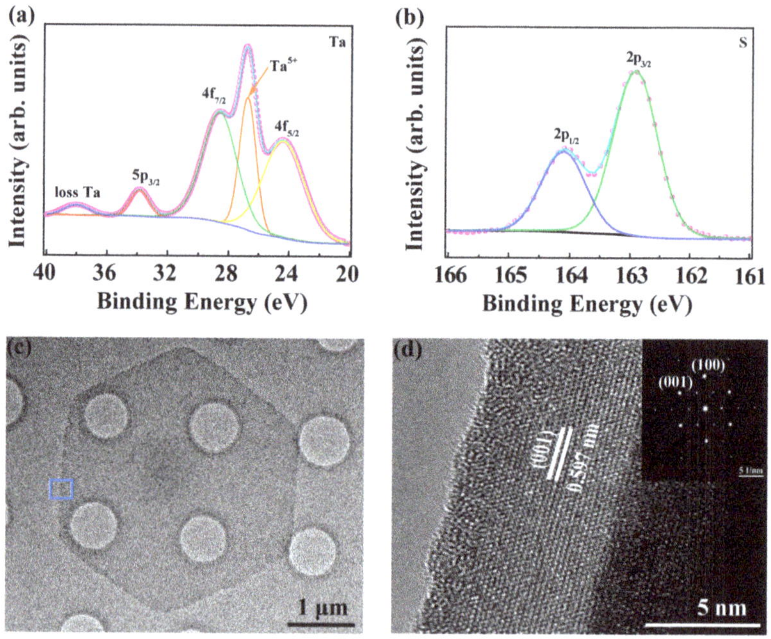

Figure 3. XPS and HRTEM measurements of 1T-TaS_2 crystal. (**a**,**b**) XPS spectra of Ta-4f, 5p and loss Ta, and S 2p orbitals. (**c**) Low-magnification image of the hexagonal shape sample. (**d**) Atomic-resolution HRTEM image of 1T-TaS_2 and the inset shows the corresponding SAED pattern.

To study the CDW phase transitions, electrical devices were fabricated. In our experiments, it is very difficult to grow large-sized thin 1T-TaS_2 crystals. Usually, the size of a 1T-TaS_2 crystal with a thickness lower than 3 nm is below 10 μm. Meanwhile, the thin 1T-TaS_2 is not so stable in the atmosphere. Hence, it is challenging to fabricate four contacts on a thin sample with a size smaller than 10 μm. To better compare the test results from samples with different thicknesses, two-contact devices were fabricated for all samples, which were proven to be valid for the resistance measurements on exfoliated samples [16]. During the electrical measurements, the positive and negative electrodes were fixed at different temperatures to prevent inconsistent modifications to hysteresis loops from Schottky effects. When the CDW phase transition occurs, the 1T-TaS_2 phase can change from an insulator to a metal, which can be identified by temperature-dependent resistance measurements. The schematics for the electrical devices are presented in Figure 4a, in which gold electrodes were fabricated on the sample by lithography termina, and more details for the procedure refer to the method. The hysteresis loops of the resistance in the heating and cooling cycles of 1T-TaS_2 with different thicknesses were measured, as shown in Figure 4b–d. In Figure 4b, the resistance of the sample with a thickness of about 50 nm decreases sharply when the temperature exceeds 160 K during heating. The resistance changing with the temperature indicates the phase transition from the CCDW phase to the NCDW phase.

On the other hand, when the temperature decreases from 300 K to 225 K, the resistance suddenly increases, indicating the occurrence of the NC–CC CDW phase transition. Similar results have been obtained from other samples with different thicknesses. In the sample with a thickness of about 20 nm in Figure 4c, the CC–NC phase transition happens at 153 K, while the NC–CC phase transition occurs at 237 K. Moreover, in the sample with a 2–3 nm thickness shown in Figure 4d, the resistance begins to decrease at 150 K during the heating process for the CC–NC phase transition, while the NC–CC phase transition temperature happens at 250 K.

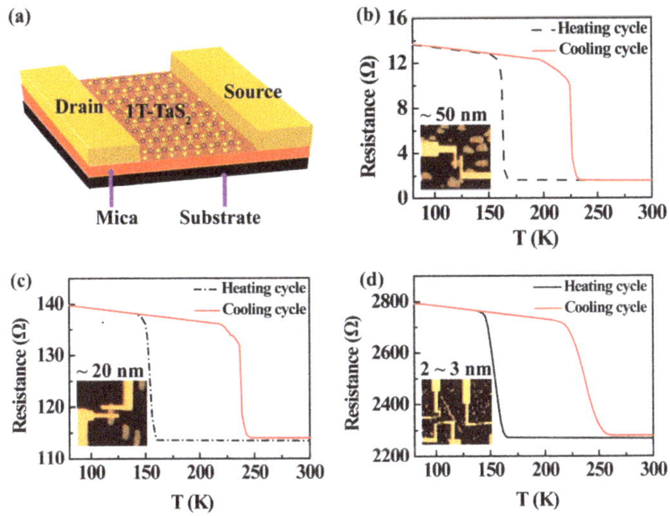

Figure 4. Temperature-dependent resistance measurements. (**a**) Schematic diagram of electrical devices. (**b–d**) Temperature-dependent resistance of samples with thicknesses of ~50 nm, ~20 nm, and 2–3 nm during the heating and cooling. The insets are microscopic photographs of electrical devices.

The above measurements were repeated more than 10 times, and the same conclusions were obtained. It is found that during the heating process, with the increase in thickness, the temperature for the CCDW-NCDW phase transition increased and ranged from 150 K to 160 K. During the cooling process, the NC-CC phase transition temperature decreased with increasing thickness, ranging from 250 K to 220 K. As the thickness decreases, the window for the hysteresis loop becomes larger, which may be attributed to the larger NC–CC phase transition barrier in thinner 1T-TaS_2 flakes due to the enhanced pinning of nucleated domain walls [10,41]. Such hysteresis loops were suggested to be caused by the domain wall propagation, leading to the ferroic performance of 2D 1T-TaS_2 crystals [16]. These ferro-rotational orders can be switched by controlling the applied voltages at a fixed temperature or the temperature at a fixed applied voltage. Therefore, thermal-driven resistance switching and the temperature of the thickness-dependent phase transition could extend the electrical and magnetic application of 1T-TaS_2.

The phase transitions were further confirmed by the temperature-dependent Raman measurements. The samples grown on mica were quickly transferred to a vacuum thermostat with a temperature range from 80 K to 260 K to collect the temperature-dependent Raman spectra. Figure 5a shows the Raman spectra of the 50 nm thick sample, indicating an obvious phase transition with emerging new peaks at 70 cm^{-1} and near 100 cm^{-1} as it warmed up. At 160 K, the peak at 70 cm^{-1} was split into two peaks at 70 cm^{-1} and 73 cm^{-1}, while the peak intensities were significantly increased near 100 cm^{-1}. Those changes are provoked by the folding of phonon modes in the Brillouin zone between the CC/NC translation [42]. Figure 5b shows the Raman spectra of the 20 nm thick sample as it warmed up. It is obvious that the peak intensities were strongly enhanced near 100 cm^{-1} at below 140 K, including the peaks at 81 cm^{-1}, 103 cm^{-1}, and 122 cm^{-1}, but no split peak was seen in the Raman spectra. Furthermore, the same measurements were performed on 2–3 nm thick samples, but no apparent phase transition was found, which may be interpreted as the instability of ultrathin TaS_2 under the irradiation of lasers. Such vanishment was also reported by previous work in TaS_2 sheets thinner than 13 nm [43,44]. The above temperature-dependent Raman measurements also confirm that the CC/NC transition temperature in thick layer samples is 140–160 K. The consistency between the resistance measurements and Raman spectra further confirms that the two-contact measurements are reasonable in our experiments.

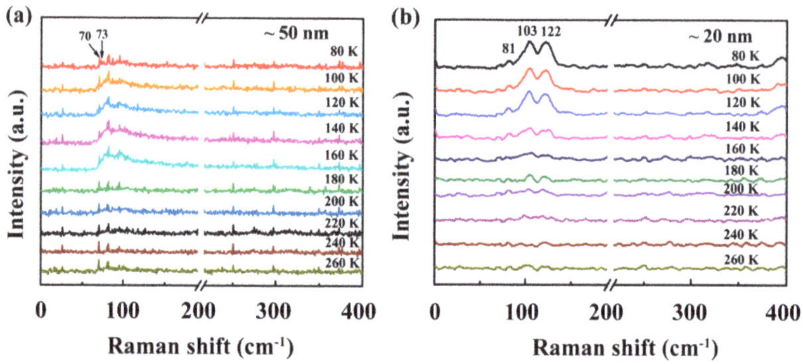

Figure 5. Raman measurements during thermal−driven phase transitions in 1T-TaS_2. (**a**,**b**) Raman characteristics of CC/NC phase transition on ~50 nm and ~20 nm samples. The Raman spectra were collected at measured temperatures rising from 80 K to 260 K.

4. Conclusions

High-quality 2D 1T-TaS_2 crystals were successfully synthesized by CVD with controllable layer numbers. The structural tests show the high quality of the grown samples. The AFM measurements demonstrated the precise control of the thickness. Based on the as-grown samples, their thickness-dependent CC/NCDW phase transitions were revealed by the combination of the temperature-dependent resistance measurements and Raman spectra. The phase transition temperature increases with increasing thickness. The transition hysteresis loops due to temperature-dependent resistance changes of 1T-TaS_2 can be used for memory devices and oscillators, which is promising for various electrical applications.

Author Contributions: Conceptualization, X.P., H.L. and H.Q.; methodology, X.P., H.L. and T.Y.; software, X.P., T.Y., H.B., J.P., L.L., F.J., H.Q. and H.L.; validation, X.P., H.L. and H.Q.; formal analysis, X.P., T.Y., H.B., J.P., L.L., F.J. and Z.H.; investigation, X.P. and H.L.; resources, X.P., H.Q., H.L. and Z.H.; data curation, X.P., T.Y., H.B., J.P., L.L., F.J. and H.L.; writing—original draft preparation, X.P., T.Y. and H.L.; writing—review and editing, X.P., H.Q. and H.L.; visualization, X.P. and H.L.; supervision, H.L.; project administration, H.L.; funding acquisition, H.L. and H.Q. All authors have read and agreed to the published version of the manuscript.

Funding: The authors are grateful to the National Natural Science Foundation of China (Nos. 52172151, 51972229, 52002275) and the National Defense Science and Technology 173 Program (2021-JCJQ-JJ-0639).

Data Availability Statement: Data available on request due to restrictions e.g., privacy or ethical.

Conflicts of Interest: The authors declare no conflict of interest.

References

1. Peng, J.; Yu, Z.; Wu, J.; Zhou, Y.; Guo, Y.; Li, Z.; Zhao, J.; Wu, C.; Xie, Y. Disorder enhanced superconductivity toward TaS_2 monolayer. *ACS Nano* **2018**, *12*, 9461–9466. [CrossRef] [PubMed]
2. Wen, W.; Zhu, Y.; Dang, C.; Chen, W.; Xie, L. Raman spectroscopic and dynamic electrical investigation of multi-state charge-wave-density phase transitions in 1T-TaS_2. *Nano Lett.* **2019**, *19*, 1805–1813. [CrossRef] [PubMed]
3. Nakata, Y.; Sugawara, K.; Chainani, A.; Oka, H.; Bao, C.H.; Zhou, S.H.; Chuang, P.Y.; Cheng, C.M.; Kawakami, T.; Saruta, Y.; et al. Robust charge-density wave strengthened by electron correlations in monolayer 1T-$TaSe_2$ and 1T-$NbSe_2$. *Nat. Commun.* **2021**, *12*, 5873. [CrossRef] [PubMed]
4. Zhao, X.; Song, P.; Wang, C.; Riis-Jensen, A.C.; Fu, W.; Deng, Y.; Wan, D.; Kang, L.; Ning, S.; Dan, J.; et al. Engineering covalently bonded 2D layered materials by self-intercalation. *Nature* **2020**, *581*, 171–177. [CrossRef]
5. Wang, P.; Huan, Y.; Yang, P.; Cheng, M.; Shi, J.; Zhang, Y. Controlled syntheses and multifunctional applications of two-dimensional metallic transition metal dichalcogenides. *Acc. Mater. Res.* **2021**, *2*, 751–763. [CrossRef]

6. Zhu, C.; Chen, Y.; Liu, F.; Zheng, S.; Li, X.; Chaturvedi, A.; Zhou, J.; Fu, Q.; He, Y.; Zeng, Q.; et al. Light-tunable 1T-TaS_2 charge-density-wave oscillators. *ACS Nano* **2018**, *12*, 11203–11210. [CrossRef]
7. Yoshida, M.; Suzuki, R.; Zhang, Y.; Nakano, M.; Iwasa, Y. Memristive phase switching in two-dimensional 1T-TaS_2 crystals. *Sci. Adv.* **2015**, *1*, e1500606. [CrossRef]
8. Afzali, M.; Mostafavi, A.; Shamspur, T. Improved perovskite solar cell with 2H–TaS_2 nanosheets as an electron transport layer using microwave irradiation. *J. Alloys Compd.* **2020**, *817*, 152742. [CrossRef]
9. Feng, Y.; Gong, S.; Du, E.; Yu, K.; Ren, J.; Wang, Z.; Zhu, Z. TaS_2 nanosheet-based ultrafast response and flexible humidity sensor for multifunctional applications. *J. Mater. Chem. C* **2019**, *7*, 9284–9292. [CrossRef]
10. Huan, Y.; Shi, J.; Zou, X.; Gong, Y.; Zhang, Z.; Li, M.; Zhao, L.; Xu, R.; Jiang, S.; Zhou, X.; et al. Vertical 1T-TaS_2 synthesis on nanoporous gold for high-performance electrocatalytic applications. *Adv. Mater.* **2018**, *30*, 1705916. [CrossRef]
11. Ekoya, B.G.M.; Shan, Y.B.; Cai, Y.C.; Okombi, N.I.; Yue, X.F.; Xu, M.S.; Cong, C.X.; Hu, L.G.; Qiu, Z.J.; Liu, R. 2H Tantalum disulfide nanosheets as substrates for ultrasensitive SERS-based sensing. *ACS Appl. Nano Mater.* **2022**, *5*, 8913–8920. [CrossRef]
12. Jarach, Y.; Rodes, L.; Ber, E.; Yalon, E.; Kanigel, A. Joule-heating induced phase transition in 1T-TaS_2 near room temperature probed by thermal imaging of power dissipation. *Appl. Phys. Lett.* **2022**, *120*, 083502. [CrossRef]
13. Jia, Y.; Liao, Y.L.; Cai, H.Z. High quality TaS_2 nanosheet SPR biosensors improved sensitivity and the experimental demonstration for the Ddetection of Hg^{2+}. *Nanomaterials* **2022**, *12*, 2075. [CrossRef]
14. Ishiguro, Y.; Suzuki, R.; Yangzhou, Z.; Kodama, N.; Takai, K. Correlation between charge density wave phase transition and hydrogen adsorption in 1T-TaS_2 thin film devices. *Nanotechnology* **2023**, *34*, 275701. [CrossRef] [PubMed]
15. Buravets, V.; Hosek, F.; Lapcak, L.; Miliutina, E.; Sajdl, P.; Elashnikov, R.; Svorcik, V.; Lyutakov, O. Beyond the platinum era-scalable preparation and electrochemical activation of TaS_2 flakes. *ACS Appl. Mater. Interfaces* **2023**, *15*, 5679–5686. [CrossRef]
16. Liu, G.; Qiu, T.; He, K.; Liu, Y.; Lin, D.; Ma, Z.; Huang, Z.; Tang, W.; Xu, J.; Watanabe, K.; et al. Electrical switching of ferro-rotational order in nanometre-thick 1T-TaS_2 crystals. *Nat. Nanotechnol.* **2023**. [CrossRef] [PubMed]
17. Sugawara, K.; Nakata, Y.; Fujii, K.; Nakayama, K.; Souma, S.; Takahashi, T.; Sato, T. Monolayer VTe_2: Incommensurate fermi surface nesting and suppression of charge density waves. *Phys. Rev. B* **2019**, *99*, 241404. [CrossRef]
18. Lian, C.-S.; Si, C.; Duan, W. Unveiling charge-density wave, superconductivity, and their competitive nature in two-dimensional $NbSe_2$. *Nano Lett.* **2018**, *18*, 2924–2929. [CrossRef]
19. Sipos, B.; Kusmartseva, A.F.; Akrap, A.; Berger, H.; Forró, L.; Tutiš, E. From mott state to superconductivity in 1T-TaS_2. *Nat. Mater.* **2008**, *7*, 960–965. [CrossRef]
20. Vaskivskyi, I.; Gospodaric, J.; Brazovskii, S.; Svetin, D.; Sutar, P.; Goreshnik, E.; Mihailovic, I.A.; Mertelj, T.; Mihailovic, D. Controlling the metal-to-insulator relaxation of the metastable hidden quantum state in 1T-TaS_2. *Sci. Adv.* **2015**, *1*, e1500168. [CrossRef]
21. Liu, G.; Rumyantsev, S.; Bloodgood, M.A.; Salguero, T.T.; Balandin, A.A. Low-frequency current fluctuations and sliding of the charge density waves in two-dimensional materials. *Nano Lett.* **2018**, *18*, 3630–3636. [CrossRef]
22. Gao, J.; Park, J.W.; Kim, K.; Song, S.K.; Park, H.R.; Lee, J.; Park, J.; Chen, F.; Luo, X.; Sun, Y.; et al. Pseudogap and weak multifractality in 2D disordered mott charge-density-wave insulator. *Nano Lett.* **2020**, *20*, 6299–6305. [CrossRef] [PubMed]
23. Bu, K.; Zhang, W.; Fei, Y.; Wu, Z.; Zheng, Y.; Gao, J.; Luo, X.; Sun, Y.-P.; Yin, Y. Possible strain induced mott gap collapse in 1T-TaS_2. *Commun. Phys.* **2019**, *2*, 146. [CrossRef]
24. Perfetti, L.; Loukakos, P.; Lisowski, M.; Bovensiepen, U.; Berger, H.; Biermann, S.; Cornaglia, P.; Georges, A.; Wolf, M. Time evolution of the electronic structure of 1T−TaS_2 through the insulator-metal transition. *Phys. Rev. Lett.* **2006**, *97*, 067402. [CrossRef]
25. Stojchevska, L.; Vaskivskyi, I.; Mertelj, T.; Kusar, P.; Svetin, D.; Brazovskii, S.; Mihailovic, D. Ultrafast switching to a stable hidden quantum state in an electronic crystal. *Science* **2014**, *344*, 177–180. [CrossRef] [PubMed]
26. Vaskivskyi, I.; Mihailovic, I.; Brazovskii, S.; Gospodaric, J.; Mertelj, T.; Svetin, D.; Sutar, P.; Mihailovic, D. Fast electronic resistance switching involving hidden charge density wave states. *Nat. Commun.* **2016**, *7*, 11442. [CrossRef]
27. Ma, L.; Ye, C.; Yu, Y.; Lu, X.F.; Niu, X.; Kim, S.; Feng, D.; Tománek, D.; Son, Y.-W.; Chen, X.H.; et al. A metallic mosaic phase and the origin of mott-insulating state in 1T-TaS_2. *Nat. Commun.* **2016**, *7*, 10956. [CrossRef] [PubMed]
28. Wu, D.; Ma, Y.; Niu, Y.; Liu, Q.; Dong, T.; Zhang, S.; Niu, J.; Zhou, H.; Wei, J.; Wang, Y.; et al. Ultrabroadband photosensitivity from visible to terahertz at room temperature. *Sci. Adv.* **2018**, *4*, eaao3057. [CrossRef] [PubMed]
29. Shi, J.; Wang, X.; Zhang, S.; Xiao, L.; Huan, Y.; Gong, Y.; Zhang, Z.; Li, Y.; Zhou, X.; Hong, M.; et al. Two-dimensional metallic tantalum disulfide as a hydrogen evolution catalyst. *Nat. Commun.* **2017**, *8*, 958. [CrossRef]
30. Pathan, M.A.K.; Gupta, A.; Vaida, M.E. Exploring the growth and oxidation of 2D-TaS_2 on Cu(111). *Nanotechnology* **2021**, *32*, 505605. [CrossRef]
31. Dombrowski, D.; Samad, A.; Murray, C.; Petrovic, M.; Ewen, P.; Michely, T.; Kralj, M.; Schwingenschlogl, U.; Busse, C. Two phases of monolayer tantalum sulfide on Au(111). *ACS Nano* **2021**, *15*, 13516–13525. [CrossRef] [PubMed]
32. Fu, W.; Chen, Y.; Lin, J.; Wang, X.; Zeng, Q.; Zhou, J.; Zheng, L.; Wang, H.; He, Y.; He, H.; et al. Controlled synthesis of atomically thin 1T-TaS_2 for tunable charge density wave phase transitions. *Chem. Mater.* **2016**, *28*, 7613–7618. [CrossRef]
33. Wang, X.; Liu, H.; Wu, J.; Lin, J.; He, W.; Wang, H.; Shi, X.; Suenaga, K.; Xie, L. Chemical growth of 1T-TaS_2 monolayer and thin films: Robust charge density wave transitions and high bolometric responsivity. *Adv. Mater.* **2018**, *30*, 1800074. [CrossRef] [PubMed]

34. Zhao, R.; Wang, Y.; Deng, D.; Luo, X.; Lu, W.J.; Sun, Y.-P.; Liu, Z.-K.; Chen, L.-Q.; Robinson, J. Tuning phase transitions in 1T-TaS_2 via the substrate. *Nano Lett.* **2017**, *17*, 3471–3477. [CrossRef]
35. Neto, A.C.; Novoselov, K. New directions in science and technology: Two-dimensional crystals. *Rep. Prog. Phys.* **2011**, *74*, 082501. [CrossRef]
36. Wang, K.; Huang, B.; Tian, M.; Ceballos, F.; Lin, M.-W.; Mahjouri-Samani, M.; Boulesbaa, A.; Puretzky, A.A.; Rouleau, C.M.; Yoon, M.; et al. Interlayer coupling in twisted WSe_2/WS_2 bilayer heterostructures revealed by optical spectroscopy. *ACS Nano* **2016**, *10*, 6612–6622. [CrossRef]
37. Duffey, J.; Kirby, R.; Coleman, R. Raman scattering from 1T-TaS_2. *Solid State Commun.* **1976**, *20*, 617–621. [CrossRef]
38. Cho, D.; Cheon, S.; Kim, K.-S.; Lee, S.-H.; Cho, Y.-H.; Cheong, S.-W.; Yeom, H.W. Nanoscale manipulation of the mott insulating state coupled to charge order in 1T-TaS_2. *Nat. Commun.* **2016**, *7*, 10453. [CrossRef]
39. Eda, G.; Yamaguchi, H.; Voiry, D.; Fujita, T.; Chen, M.; Chhowalla, M. Photoluminescence from chemically exfoliated MoS_2. *Nano Lett.* **2011**, *11*, 5111–5116. [CrossRef]
40. Cain, J.D.; Oh, S.; Azizi, A.; Stonemeyer, S.; Dogan, M.; Thiel, M.; Ercius, P.; Cohen, M.L.; Zettl, A. Ultranarrow TaS_2 Nanoribbons. *Nano Lett.* **2021**, *21*, 3211–3217. [CrossRef]
41. He, R.; Okamoto, J.; Ye, Z.; Ye, G.; Anderson, H.; Dai, X.; Wu, X.; Hu, J.; Liu, Y.; Lu, W.; et al. Distinct surface and bulk charge density waves in ultrathin 1T−TaS_2. *Phys. Rev. B* **2016**, *94*, 201108. [CrossRef]
42. Grisafe, B.; Zhao, R.; Ghosh, R.K.; Robinson, J.A.; Datta, S. Electrically triggered insulator-to-metal phase transition in two-dimensional (2D) heterostructures. *Appl. Phys. Lett.* **2018**, *113*, 142101. [CrossRef]
43. Yu, Y.; Yang, F.; Lu, X.F.; Yan, Y.J.; Cho, Y.-H.; Ma, L.; Niu, X.; Kim, S.; Son, Y.-W.; Feng, D.; et al. Gate-tunable phase transitions in thin flakes of 1T-TaS_2. *Nat. Nanotechnol.* **2015**, *10*, 270–276. [CrossRef] [PubMed]
44. Darancet, P.; Millis, A.J.; Marianetti, C.A. Three-dimensional metallic and two-dimensional insulating behavior in octahedral tantalum dichalcogenides. *Phys. Rev. B* **2014**, *90*, 045134. [CrossRef]

nanomaterials

Article

Colossal Magnetoresistance in Layered Diluted Magnetic Semiconductor Rb(Zn,Li,Mn)$_4$As$_3$ Single Crystals

Yi Peng [1,2], Luchuan Shi [1,2], Guoqiang Zhao [1], Jun Zhang [1], Jianfa Zhao [1], Xiancheng Wang [1], Zheng Deng [1,2,*] and Changqing Jin [1,2,*]

1 Institute of Physics, Chinese Academy of Sciences, Beijing 100190, China; ypeng@iphy.ac.cn (Y.P.); g.q.zhao@iphy.ac.cn (G.Z.); zhaojf@iphy.ac.cn (J.Z.)
2 School of Physics, University of Chinese Academy of Sciences, Beijing 101408, China
* Correspondence: dengzheng@iphy.ac.cn (Z.D.); jin@iphy.ac.cn (C.J.)

Abstract: Diluted magnetic semiconductors (DMSs) with tunable ferromagnetism are among the most promising materials for fabricating spintronic devices. Some DMS systems have sizeable magnetoresistances that can further extend their applications. Here, we report a new DMS Rb(Zn$_{1-x-y}$Li$_y$Mn$_x$)$_4$As$_3$ with a quasi-two-dimensional structure showing sizeable anisotropies in its ferromagnetism and transverse magnetoresistance (MR). With proper charge and spin doping, single crystals of the DMS display Curie temperatures up to 24 K. Analysis of the critical behavior via Arrott plots confirms the long-range ferromagnetic ordering in the Rb(Zn$_{1-x-y}$Li$_y$Mn$_x$)$_4$As$_3$ single crystals. We observed remarkable intrinsic MR effects in the single crystals (i.e., a positive MR of 85% at 0.4 T and a colossal negative MR of −93% at 7 T).

Keywords: diluted magnetic semiconductor; colossal magnetoresistance; quasi-two-dimensional structure; single crystal

check for updates

Citation: Peng, Y.; Shi, L.; Zhao, G.; Zhang, J.; Zhao, J.; Wang, X.; Deng, Z.; Jin, C. Colossal Magnetoresistance in Layered Diluted Magnetic Semiconductor Rb(Zn,Li,Mn)$_4$As$_3$ Single Crystals. *Nanomaterials* **2024**, *14*, 263. https://doi.org/10.3390/nano14030263

Academic Editor: Filippo Giannazzo

Received: 12 December 2023
Revised: 10 January 2024
Accepted: 15 January 2024
Published: 25 January 2024

1. Introduction

Diluted magnetic semiconductors (DMSs) have received extensive attention due to their significant potential for spintronic applications [1–3]. The intriguing properties of DMSs include their ability to mediate ferromagnetic interactions by tuning the conduction carriers and the strong coupling between the carriers and local spins [4,5]. The former can provide novel opportunities to control ferromagnetism, and the latter can lead to galvanomagnetic properties (e.g., magnetoresistance (MR) and anisotropic magnetoresistance effects) [6–9]. The most well-known examples of substantial magnetoresistance are magnetic multilayers and manganites, in which the coupling between the spin and charge is essential [10,11]. Similar coupling has generated anomalous negative MR in many ferromagnetic DMS materials (e.g., III-V based (Ga,Mn)As and II-II-V based (Ba,K)(ZnMn)$_2$As$_2$) [12,13]. However, most of the reported MR in DMS single crystals or single-phase thin films are relatively weak, with magnitudes of 10–30%.

The aforementioned (Ba,K)(ZnMn)$_2$As$_2$ has a layered structure to allow for the spatial and electronic separation of charge and spin doping [14]. It belongs to a new generation of DMSs in which the charges (or carriers) and spins are doped independently. The initial motivation to design and synthesize these new DMSs was to overcome the nonequilibrium Mn doping in the classical III-V based DMSs (e.g., (Ga,Mn)As, (In,Mn)Sb, etc.) [15]. This difficulty leads to limited solid solutions of Mn^{2+} in III-V-based DMSs. Consequently, the specimens of these DMSs are chemically metastable [16]. Ferromagnetic films are available only as thin films, and their material quality exhibits high sensitivity in preparation methods and annealing treatments. Eventually, the improvement in their Curie temperatures (T_C) was prohibitive. Moreover, Mn substitution provided hole carriers together with local spins. Thus, one could not conduct electron doping to obtain n-type materials or p-n junctions for spintronic devices [17].

Taking advantage of the separated charge and spin doping, the (Ba,K)$(ZnMn)_2As_2$ achieved a reliable T_C of 230 K, which is close to room temperature [18]. Further studies stated that external compression on (Ba,K)$(ZnMn)_2As_2$ reduced the distance between the ZnAs layers and then decreased its T_C [19]. In other words, extending the distance of the ZnAs layers could result in a higher T_C. $RbZn_4As_3$ has a well-defined quasi-two-dimensional layered structure with a larger ZnAs-layer distance (4.3 Å) than that of (Ba,K)$(ZnMn)_2As_2$ (3.7 Å). Although we did not obtain a high T_C after magnetic and charge doping, interesting galvanomagnetic properties were discovered. In this work, we report $Rb(Zn_{1-x-y}Li_yMn_x)_4As_3$ as a new type of DMS. We found remarkable MR effects on the single crystals, thus ruling out the possible influence of the grain boundaries in polycrystalline samples [20,21]. This DMS has a layered structure as the parent phase, which consists of infinitely stacked Rb layers and ZnAs layers. A detailed and comprehensive analysis of the critical behavior yielded the critical exponents β, γ, and δ. These exponents suggest that the mean-field model is the applicable theoretical model, indicating the long-range ferromagnetism in this material. Owing to the distinct anisotropic structure, sizeable anisotropies were observed in the ferromagnetism and transverse MR. Moreover, large positive MRs at low fields and colossal negative MRs at high fields exist in single crystals. It is worth noting that although large MRs were found in polycrystalline (Sr,K)$(Zn,Mn)_2As_2$ and (Ba,K)$(Cd,Mn)_2As_2$, one cannot rule out the contribution from the grain boundaries, which diffusely exist in the polycrystalline samples. These intriguing features of $Rb(Zn_{1-x-y}Li_yMn_x)_4As_3$ should benefit applications such as memory devices and magnetic sensors [3,22,23].

2. Materials and Methods

Polycrystalline specimens of $Rb(Zn_{1-x-y}Li_yMn_x)_4As_3$ were prepared via the solid-state reaction method with high-purity pristine reagents [17,24]. Firstly, Rb_3As and Li_3As were synthesized as precursors from Rb grains (99.9%), Li grains (99.9%), and As powder (99.99%). To avoid the oxidization and corrosion of the alkali metals (Rb and Li), raw materials were put directly into titanium tubes and sealed under an argon atmosphere at 1 bar. After that, the mixtures were heated slowly to 350 °C and then held for 5 h due to the low melting point and activity of the alkali metals. Afterwards, Zn powder (99.99%), Mn powder (99.99%), As powder, and the two precursors were mixed well and ground at the intended stoichiometric ratios. The mixtures were sealed in titanium tubes under an argon atmosphere after pelletization. The tubes were sealed in quartz ampoules before being sintered at 700 °C for 20 h. The collected materials were ground and pelletized before being sintered at 600 °C for another 20 h. This annealing process was necessary to complete the reaction and produce high-purity samples.

The growth of the single crystals was performed utilizing the self-flux method. High-purity elements, Rb grains, Zn powder, Li grains, Mn powder, and As powder were mixed at a ratio of 1:3.8$-y$:y:0.2:3. The mixture was then loaded in alumina crucibles, which were sealed in tantalum tubes and quartz ampoules. The ampoules were heated up to 1000 °C for 2 h and then quenched in ice water or liquid nitrogen after cooling the furnace temperature down to 930 °C. Note that all the processes were performed under the protection of high-purity argon. The entirety of a single crystal, after quenching, could be easily cleaved along its c-axis due to the quasi-two-dimensional layered structure of the titled material. Generally, the width of a perfect single crystal was only limited to the dimensions of the alumina crucible, and thus, the size of an available single crystal could be up to 10 mm × 7 mm × 1 mm.

Powder X-ray diffraction (XRD) was conducted with a Philips X'pert diffractometer (Malvern Panalytical Ltd, Malvern, UK) at room temperature to analyze the phase purity and structural parameters. Cu-Kα radiation and 2θ scanning with a range from 10° to 120° were used during the XRD measurement. Energy dispersive X-ray analysis (EDX) with a commercial scanning electron microscope (SEM, Hitachi High-Tech Science Co., Ltd., Tokyo, Japan) was used to analyze chemical compositions of single crystals. Thus, the real atom ratios of single crystal samples are used in the following sections. In addition, considering the possible inhomogeneity of polycrystalline samples, we established the doping contents as the nominal ratio.

The characterization of *dc* magnetic susceptibility for all the samples was accomplished using a superconducting quantum interference device (SQUID, Quantum design, San Diego, CA, USA). The measuring temperatures were from 2 K to 300 K, and the measuring fields were up to 7 T. Quartz holders and brass holders with two quartz cylinders were used for the in- and out-of-*ab* plane (i.e., external field H parallel to the *ab*-plane and *c*-axis separately) magnetic measurement of single crystals, respectively. To ensure that the samples measured in the Arrott plot were initially magnetized, the isothermal magnetizations were measured after the samples were warmed up well above T_C.

Electricity transport measurements were conducted with a physical property measurement system (PPMS, Quantum Design, San Diego, CA, USA). Similarly, the measuring temperatures were from 2 K to 300 K, and the measuring fields were up to 14 T. The single crystal samples were cleaved to obtain a clean, fresh surface for good ohmic contact. A standard four-wire method was employed to eliminate contact resistance with silver paint as an electrical contact and Pt wires as electrical leads. A current of 0.1 mA was used during all transport measurements. For Hall effect measurements, the layered $Rb(Zn_{1-x-y}Li_yMn_x)_4As_3$ single crystals were cut into thin flakes with a typical size of 5 mm × 1 mm × 0.05 mm. Below 10 K, Hall resistance was difficult to measure due to the oversized magnetoresistance from slightly asymmetric Hall contacts.

3. Results

3.1. Crystal Structure

The parent phase $RbZn_4As_3$ crystallizes into the quasi-two-dimensional tetragonal structure with the space group *P4/mmm* (No. 123, Z = 1), as shown in Figure 1a [25]. The structure is closely related to the β-$BaZn_2As_2$ or $BaFe_2As_2$ superconductor with a typical $ThCr_2Si_2$-type structure [26]. Similar to β-$BaZn_2As_2$, the lattice of $RbZn_4As_3$ consists of infinitely stacked Rb layers and ZnAs layers, which are based on edge-shared $ZnAs_4$ tetrahedra. The double-stacked $ZnAs_4$ tetrahedra along the *c*-axis distinguish $RbZn_4As_3$ from β-$BaZn_2As_2$.

Isovalent Zn^{2+}/Mn^{2+} substitution was used to provide local spins, while Li^+ was doped at the Zn^{2+} site for itinerate carriers. The nominal concentrations of Mn and Li can reach 20% in polycrystalline samples. Within this doping level, all of the peaks in the XRD patterns of polycrystalline samples with distinct Li and Mn doping levels can be indexed with a $RbZn_4As_3$ structure, suggesting that they share the same structure as the parent phase. To obtain the lattice constants a and c, Rietveld refinements were performed with GSAS software [27]. The refinement of the $Rb(Zn_{0.85}Li_{0.1}Mn_{0.05})_4As_3$ polycrystalline sample is plotted in Figure 1b as an example. According to a series of XRD patterns from varying Li- and Mn-doping polycrystalline samples, we successively obtained the cell volumes (V) of $Rb(Zn_{0.9-x}Li_{0.1}Mn_x)_4As_3$, which are 180.85 ($x$ = 0.05), 181.04 (x = 0.1), 181.05 (x = 0.15), and 181.51 (x = 0.2) Å^3, while the V values of $Rb(Zn_{0.9-y}Li_yMn_{0.1})_4As_3$ are 181.27 (y = 0.05), 181.04 (y = 0.1), 180.96 (y = 0.15), and 180.78 (y = 0.2) Å^3. The results are plotted in Figure 1c, which demonstrates that the value of V monotonically increases with nominal Mn concentrations but decreases with nominal Li-doping levels, indicating successful chemical substitution. The XRD pattern of $Rb(Zn_{0.83}Li_{0.1}Mn_{0.07})_4As_3$ single crystals is shown in Figure 1d. Owing to the layered structure, all of the crystals are sheet-like and grow along the crystallographic *c*-axis. Thus, only the (0 0 l) peaks appear with 2θ scanning. As shown in the inset, the size of a typical sheet is around 2 × 1 × 0.02 mm^3.

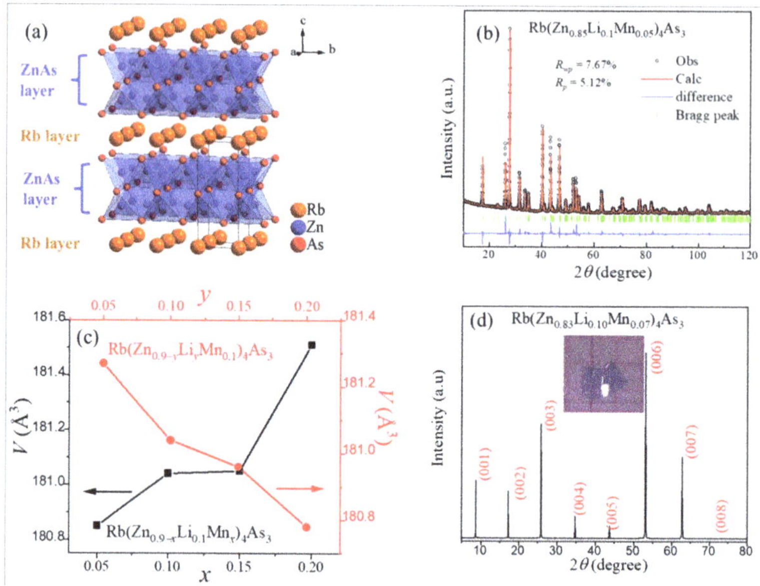

Figure 1. (**a**) Crystal structure of $RbZn_4As_3$ with stacking Rb layers and ZnAs layers. The black frame shows the unit cell. (**b**) The XRD pattern and the corresponding Rietveld refinement of $Rb(Zn_{0.85}Li_{0.1}Mn_{0.05})_4As_3$. (**c**) Lattice constants versus doping levels for $Rb(Zn_{0.9-x}Li_{0.1}Mn_x)_4As_3$ and $Rb(Zn_{0.9-y}Li_yMn_{0.1})_4As_3$, respectively. (**d**) The XRD pattern $Rb(Zn_{0.83}Li_{0.1}Mn_{0.07})_4As_3$ single crystal.

3.2. Magnetic Properties

For polycrystalline samples, the series of $Rb(Zn_{0.95-y}Li_yMn_{0.05})_4As_3$ samples have the most significant magnetizations. Thus, we focus our discussion on these samples in the following sections. Figure 2a shows the temperature-dependent magnetization M(T) under zero-field-cooling (ZFC) and field-cooling (FC) processes of $Rb(Zn_{0.95-y}Li_yMn_{0.05})_4As_3$ at an external field of 500 Oe. $Rb(Zn_{0.90}Li_{0.05}Mn_{0.05})_4As_3$ is nearly paramagnetic with a temperature down to 5 K. Correspondingly, the field-dependent magnetization M(H) of $Rb(Zn_{0.90}Li_{0.05}Mn_{0.05})_4As_3$ is a closed loop at 5 K. The slight S-shape of the loop indicates the presence of a short-range ferromagnetic correlation. More Li-doping induces stronger ferromagnetic-like behaviors. M(T) curves show T_C values of about 18 and 25 K for $Rb(Zn_{0.85}Li_{0.10}Mn_{0.05})_4As_3$ and $Rb(Zn_{0.80}Li_{0.15}Mn_{0.05})_4As_3$, respectively. With further increasing Li doping level to 0.2, the T_C slightly decreases to about 19 K. However, it is worth noting that the magnitude of magnetization of $Rb(Zn_{0.80}Li_{0.2}Mn_{0.05})_4As_3$ is much smaller than that of $Rb(Zn_{0.80}Li_{0.15}Mn_{0.05})_4As_3$. Furthermore, at 5 K, the hysteresis loop of $Rb(Zn_{0.80}Li_{0.2}Mn_{0.05})_4As_3$ becomes thinner than that of $Rb(Zn_{0.80}Li_{0.15}Mn_{0.05})_4As_3$, as shown in Figure 2b, indicating that an excessive doping level of Li damages the ferromagnetic ordering in the titled DMS materials. Notwithstanding its weak ferromagnetic interaction, it is significant for the contrast of the $M(T)$ curve with a Li-free sample, which evidently displays paramagnetic behavior, as shown in Figure 2c. In short, the evolution of ferromagnetism with the Li doping level suggests that the itinerant carriers offered by the heterovalent substitution Zn^{2+}/Li^{+} induce ferromagnetism.

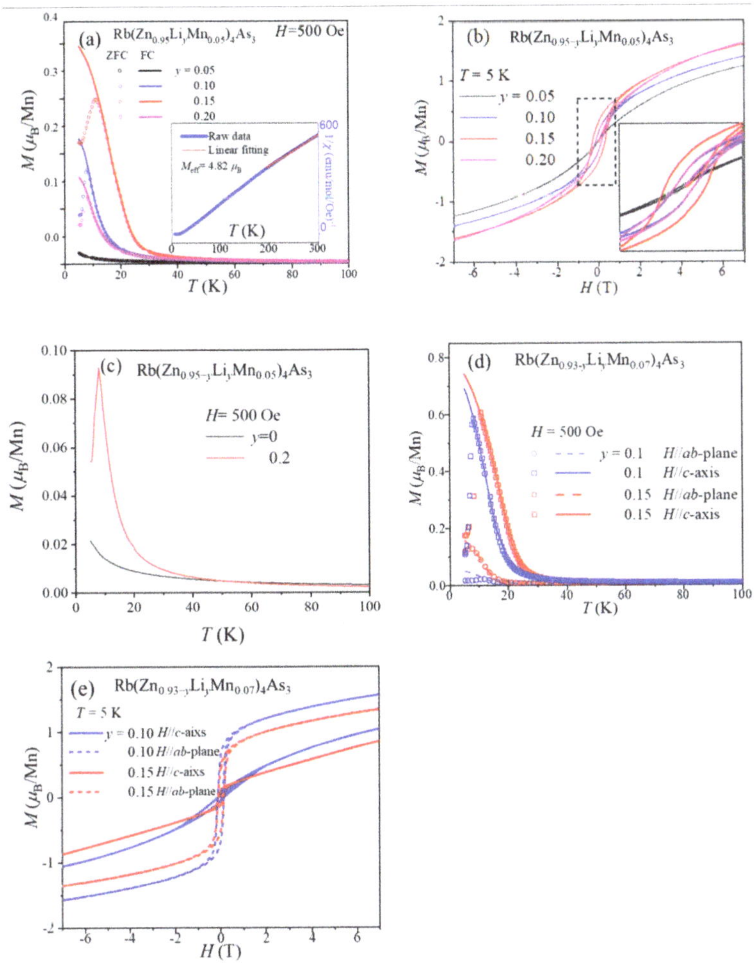

Figure 2. (**a**) Temperature-dependent magnetization of polycrystalline Rb(Zn$_{0.95-y}$Li$_y$Mn$_{0.05}$)$_4$As$_3$ with y = 0.05, 0.10, 0.15, and 0.20 after ZFC and FC processes under an external field of 500 Oe. The inset is the temperature-dependent reciprocal of magnetic susceptibility and corresponding linear fitting. (**b**) Field-dependent magnetization of Rb(Zn$_{0.95-y}$Li$_y$Mn$_{0.05}$)$_4$As$_3$ at 5 K. The inset is the enlarged magnetic loops at low fields. (**c**) M(T) curves of Rb(Zn$_{0.95}$Mn$_{0.05}$)$_4$As$_3$ and Rb(Zn$_{0.75}$Li$_{0.10}$Mn$_{0.05}$)$_4$As$_3$ in the ZFC process. (**d**) M(T) curves of Rb(Zn$_{0.83}$Li$_{0.10}$Mn$_{0.07}$)$_4$As$_3$ and Rb(Zn$_{0.78}$Li$_{0.15}$Mn$_{0.07}$)$_4$As$_3$ single crystals with an external field parallel to the c-axis and ab-plane. (**e**) M(H) curves of Rb(Zn$_{0.83}$Li$_{0.10}$Mn$_{0.07}$)$_4$As$_3$ and Rb(Zn$_{0.78}$Li$_{0.15}$Mn$_{0.07}$)$_4$As$_3$ single crystals with an external field parallel to the c-axis and ab-plane at 5 K.

The temperature-dependent inverse susceptibility ($1/\chi(T)$) and corresponding Curie-Weiss fitting of Rb(Zn$_{0.85}$Li$_{0.10}$Mn$_{0.05}$)$_4$As$_3$ are plotted in the inset of Figure 2a as a typical example. On the basis of high-temperature fitting, the effective magnetic moments (M_{eff}) obtained via Curie-Weiss fitting ($1/\chi = (T - \theta)/C$) of the paramagnetic region are around 5 μ_B/Mn [28]. The M(H) curves of the above three samples display open loops of ferromagnetism, as shown in Figure 2b. The coercive fields are about 0.2, 0.4, and 0.2 T, respectively.

Similar Li-doping-dependent behaviors can be found in single-crystal samples. Thus, the M(T) and M(H) curves of Rb(Zn$_{0.83}$Li$_{0.10}$Mn$_{0.07}$)$_4$As$_3$ and Rb(Zn$_{0.78}$Li$_{0.15}$Mn$_{0.07}$)$_4$As$_3$, which show robust ferromagnetism, are plotted in Figure 2d,e as typical examples. Distinct magnetic anisotropy can be found in both the M(T) and M(H) curves. It is clear that the easy axis of magnetization is along the c-axis. In Figure 2e, the saturation magnetizations (M_{sat}) along the c-axis are 1.3 and 1.1 μ_B/Mn for $Rb(Zn_{0.83}Li_{0.10}Mn_{0.07})_4As_3$ and $Rb(Zn_{0.78}Li_{0.15}Mn_{0.07})_4As_3$, respectively. On the other hand, the M_{sat} in the ab-plane is about 0.2 μ_B/Mn. Both the M_{sat} and M_{eff} are comparable to those of (Ga,Mn)As and $(Ba,K)(ZnMn)_2As_2$ [29,30].

3.3. *Magnetic Critical Behaviors*

Thanks to homogeneous single crystals, we could obtain more accurate magnetization data in the vicinity of the T_C to analyze the critical behavior and calculate corresponding critical exponents. By analyzing the values of critical exponents, ferromagnetic interaction in $Rb(Zn_{1-x-y}Li_yMn_x)_4As_3$ can be determined. Firstly, the determination of T_C is indispensable and crucial. To obtain the precise T_C, we analyzed the critical behaviors of two samples with the mean-field-behavior Arrott plot method as follows:

$$H/M = at + bM^2. \tag{1}$$

where a and b are constants and $t = (T_C - T)/T_C$ is the reduced temperature with an absolute value of $|t| \ll 1$. In the Arrott plot, all the curves form a series of parallel lines in the high-field region. T_C can be determined when the intercept of the parallel line becomes zero [31–33]. The obtained T_C values are 20 and 24 K for $Rb(Zn_{0.83}Li_{0.10}Mn_{0.07})_4As_3$ and $Rb(Zn_{0.78}Li_{0.15}Mn_{0.07})_4As_3$ single crystals, respectively. Figure 3a–c are the Arrott plots of $Rb(Zn_{0.83}Li_{0.10}Mn_{0.07})_4As_3$ single crystals in the vicinity of the T_C. It is worth noting that a short-range magnetic transition, like a spin-glass transition, cannot have a positive intercept in the Arrott plot. On the other hand, when we considered those modified Arrott plots in the situation of short-range exchange interaction, e.g., a three-dimensional Heisenberg model or a three-dimensional Ising model, the results markedly indicate a much lower T_C (about 7 K) according to the generalized equation of state:

$$(H/M)^{1/\gamma} = at + bM^{1/\beta}, \tag{2}$$

where β and γ are critical exponents ($\beta = 0.365$ and $\gamma = 1.386$ for the 3D Heisenberg model; $\beta = 0.325$ and $\gamma = 1.24$ for the 3D Ising model; $\beta = 0.5$ and $\gamma = 1$ for the long-range mean-field model), as shown in Figure 3b,c, respectively, whereas the preceding mean-field model is much more consistent with the minimum of the derivative of magnetization $dM(T)/dT$ in Figure 3d. For these three models, the high-field parts of the ideal Arrott plots are supposed to be parallel. In other words, they should show the temperature-independent slopes K in the high-field region. Thus, we defined the relative variation in slope $\Delta K = (K(T)/K(T_C) - 1)$. Here, the slope $K(T)$ is apparently temperature-dependent. Hence, the distribution of ΔK can effectively indicate which model is the most practical for samples to decide the critical exponents and determine the ferromagnetic interaction. Figure 3e shows the temperature dependence of the ΔK distribution. Obviously, the mean-field model possesses the most concentrated distribution at a value of 0, and the other two models have more dispersed distributions. This suggests that the mean-field model is the most suitable one.

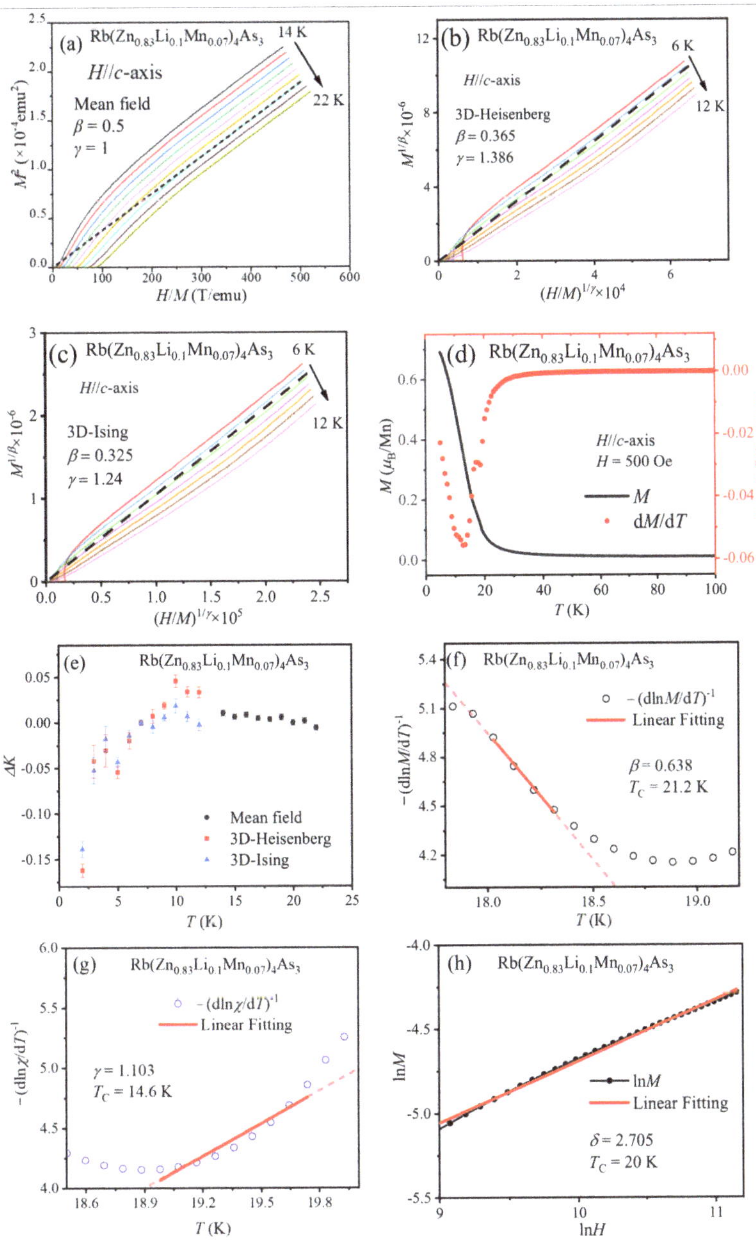

Figure 3. (**a**) Arrott plot of $Rb(Zn_{0.83}Li_{0.10}Mn_{0.07})_4As_3$ single crystals between 14 and 22 K with a step of 1 K for the mean-field model. (**b**) Arrott plot of $Rb(Zn_{0.83}Li_{0.10}Mn_{0.07})_4As_3$ single crystals between 6 and 12 K with a step of 1 K for the 3D Heisenberg model. (**c**) The 3D Ising model. (**d**) M(T) and dM(T)/dT curves of $Rb(Zn_{0.83}Li_{0.10}Mn_{0.07})_4As_3$ single crystals with an external field parallel to the c-axis. (**e**) The temperature-related distribution of the relative variation in slope from linear fitting equations of high-field Arrott plots. (**f**) Kouvel–Fisher plot of $Rb(Zn_{0.83}Li_{0.10}Mn_{0.07})_4As_3$ single crystals and corresponding linear fitting for $t > 0$. (**g**) Kouvel–Fisher plot of $Rb(Zn_{0.83}Li_{0.10}Mn_{0.07})_4As_3$ single crystals and corresponding linear fitting for $t < 0$. (**h**) Log-log plot at 20 K and corresponding linear fitting obtained from the M(H) plot at 20 K.

Additionally, Kouvel–Fisher plots, another feasible and convenient analysis method to obtain relevant parameters in the vicinity of the critical temperature, were used to provide further evidence for our results [34,35]. On the basis of the critical equation of state, there are the following power law relations among magnetization M, susceptibility χ, reduced temperature t (or Curie temperature T_C), and critical exponents:

$$M \propto t^{\beta}, \text{ when } H = 0 \text{ and } t > 0; \tag{3}$$

$$\chi \propto (-t)^{-\gamma}, \text{ when } H \rightarrow 0 \text{ and } t < 0; \tag{4}$$

$$M \propto H^{1/\delta}, \text{ when } H \rightarrow 0 \text{ and } t = 0, \tag{5}$$

where $\delta = 1 + \gamma/\beta$ according to the critical scaling analysis. After applying a logarithm and differential to Equations (3) and (4), one can draw and fit the linear relations of $-(\mathrm{dln}M/\mathrm{d}T)^{-1}$ versus T and $-(\mathrm{dln}\chi/\mathrm{d}T)^{-1}$ versus T in the critical region, i.e., Kouvel–Fisher plots, as

$$(\mathrm{dln}M/\mathrm{d}T)^{-1} = -(T_C - T)/\beta \tag{6}$$

and

$$(\mathrm{dln}M/\mathrm{d}T)^{-1} = (T_C - T)/\gamma, \tag{7}$$

in which the magnitude of the reciprocal of slope represents β and γ, respectively, and the horizontal intercept is T_C. For the same sample, the $Rb(Zn_{0.73}Li_{0.1}Mn_{0.07})_4As_3$ single crystal, its Kouvel–Fisher plots are shown in Figure 3e,f. As shown in the two figures, in the narrow enough range of $|t|$, the fitting β and γ are 0.638 and 1.103, and the corresponding T_C's are 21.2 and 14.6 K, respectively. In order to guarantee that the fitting is carried out in the vicinity of the transition region, the selected data points for linear fitting must meet the requirement of $|t| < 0.3$. Despite the deviation of T_C, the fitting results are closer to the former analysis based on the classic Arrott plot method. Meanwhile, it is feasible to determine another critical exponent, δ, by acquiring the $M(H)$ data at T_C and plotting the corresponding log-log plot according to Equation (5). As shown in Figure 3h, the linear fitting gives $\delta = 2.705$, in agreement with the value $\delta = 1 + \gamma/\beta = 2.729$ from the obtained β and γ. Thus, our discussions about critical behavior are self-consistent, and the results confirm the long-range ferromagnetism of the single crystals.

3.4. Magnetoresistance and Hall Effect

The temperature dependence of resistivity ($\rho(T)$) of the aforementioned samples was measured. All of the samples exhibit semiconducting behavior, namely an increase in ρ with decreasing temperature. The conductivity is enhanced with increasing Li concentrations, which act as charge doping. As typical examples, Figure 2a shows the $\rho(T)$ curves of ferromagnetic $Rb(Zn_{0.83}Li_{0.10}Mn_{0.07})_4As_3$ and $Rb(Zn_{0.78}Li_{0.15}Mn_{0.07})_4As_3$ single crystals. The decrease in resistivity is pronounced at low temperatures. Similar behaviors have been reported in other DMS materials, e.g., (Ga,Mn)As and Li(Zn,Mn)As [17,36]. $\ln\rho$ versus $1/T$ of $Rb(Zn_{0.83}Li_{0.10}Mn_{0.07})_4As_3$ single crystals is shown in Figure 4a. The band gap E_g ~19 meV is calculated from active model fitting:

$$\rho = \rho_0 \exp(E_g/2k_B T). \tag{8}$$

This is only one-tenth of the parent compound (0.2 eV), which also indicates successful charge doping by Li substitutions.

Sizeable transverse magnetoresistance (fields perpendicular to current) can be found in ferromagnetic samples at low temperatures, ruling out the possible influence of grain boundaries in polycrystalline samples. The MR ratio is defined as MR $\equiv (\rho(H) - \rho(0\mathrm{T}))/\rho(0\mathrm{T}) = \Delta\rho/\rho(0\mathrm{T})$ [37]. Figure 4b shows this giant MR in the samples $Rb(Zn_{0.93-y}Li_yMn_{0.07})_4As_3$, where y = 0.05, 0.1, 0.15, 0.2, among which the doping level of 10% Li contributes to the highest negative MR up to −93% at 7 T and the highest positive MR up to 85% at 0.4 T simultaneously. Given that MR changes slightly when the external field exceeds 7 T, the following measurements were only conducted within ±7 T. Figure 4c shows the temperature-dependent MR of $Rb(Zn_{0.83}Li_{0.1}Mn_{0.07})_4As_3$ single crystals along the *c*-axis and *ab*-plane. With the deceasing temperature, colossal MR appears in the vicinity of the Curie temperature T_C and enlarges monotonically. Meanwhile, the noticeable anisotropy of transverse MR is consistent with the magnetic anisotropy in Figure 2d,e. This indicates that the MR effect of ferromagnetic $Rb(Zn_{0.83}Li_{0.10}Mn_{0.07})_4As_3$ originates from a reduction in the carrier-spin scattering when local spins tend to be arranged completely parallel under external fields.

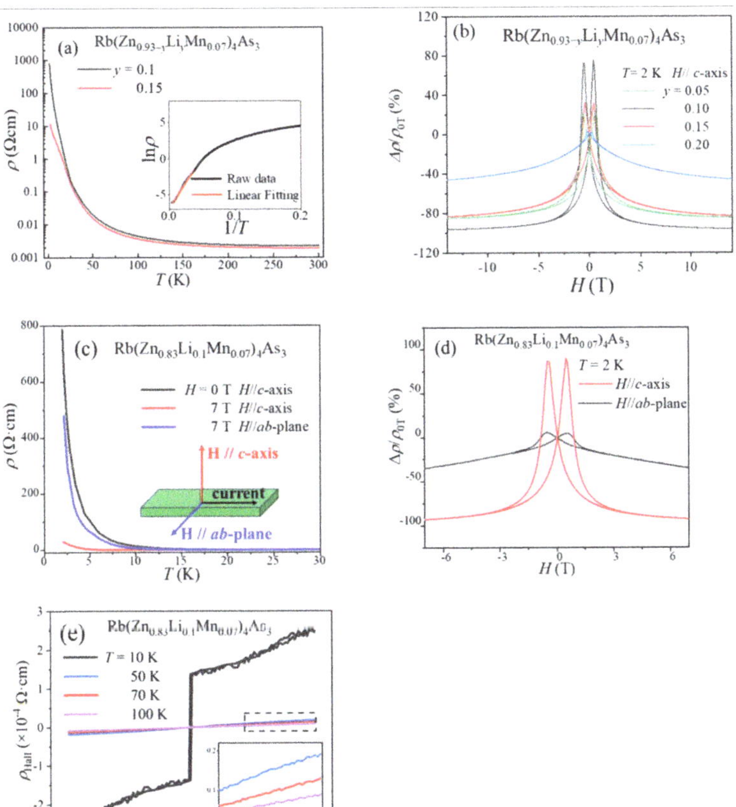

Figure 4. (**a**) Temperature-dependent resistivity of $Rb(Zn_{0.83}Li_{0.10}Mn_{0.07})_4As_3$ and $Rb(Zn_{0.78}Li_{0.15}Mn_{0.07})_4As_3$ single crystals. The inset is the active model fitting of $Rb(Zn_{0.83}Li_{0.10}Mn_{0.07})_4As_3$. (**b**) Transverse magnetoresistance of $Rb(Zn_{0.93-y}Li_yMn_{0.07})_4As_3$ single crystals with H of 14 T parallel to the *c*-axis. (**c**) ρ(T) of $Rb(Zn_{0.83}Li_{0.10}Mn_{0.07})_4As_3$ single crystals with H of 7 T parallel to the *c*-axis and *ab*-plane at low temperatures. The inset is the configuration of the external fields and measured current. (**d**) Transverse magnetoresistance of the $Rb(Zn_{0.83}Li_{0.10}Mn_{0.07})_4As_3$ single crystals with H parallel to the *c*-axis and *ab*-plane at 2 K. (**e**) Hall effect measurement of $Rb(Zn_{0.83}Li_{0.10}Mn_{0.07})_4As_3$ single crystals. The inset is the enlarged high-field region.

Figure 4d shows the field-dependent MR of $Rb(Zn_{0.83}Li_{0.1}Mn_{0.07})_4As_3$ single crystals at 2 K. With an external field along the *c*-axis, it displays a positive MR of 85% at 0.4 T and a negative MR of −93% at 7 T. It is worth noting that a positive MR with such a large magnitude has rarely been reported in previous DMS materials. In contrast, when using the $H//ab$-plane, the maximum values of the positive and negative MR become 7% and −35%, respectively. The saturation status of MR is consistent with the hysteresis loops. Magnetization and MR saturate at about 3 T along the easiest axis, namely the *c*-axis, while neither of them saturates even at 7 T. The positive MR can be attributed to the delayed rotation of spins under external fields, which corresponds to coercive fields (Figure 2e) [13]. Such a remarkable contrast of MR of low and high fields can stand for an off-and-on status and act as a magnetic switch in a circuit. Meanwhile, the large anisotropy of transverse MR makes the titled DMS a good candidate for magnetic field sensors [38].

To determine the carrier type and concentration, the Hall effect was measured. Figure 4e is the field-dependent Hall resistance ($\rho_{Hall}(H)$) of $Rb(Zn_{0.83}Li_{0.1}Mn_{0.07})_4As_3$ single crystals below and above T_C. Above T_C, $\rho_{Hall}(H)$ curves are linear, and the carrier type can be easily determined as the *p*-type. Below T_C, the $\rho_{Hall}(H)$ of a ferromagnetic conductor is expressed as

$$R_{Hall}(\mathrm{H}) = R_0 H + R_s M, \tag{9}$$

where R_0 is the ordinary Hall coefficient, R_s denotes the anomalous Hall coefficient, and M denotes the magnetization of the samples [39,40]. At 10 K, the anomalous Hall component dominates the low-field region, so the carrier concentration is calculated with high-field data where $\rho_{Hall}(\mathrm{H})$ curves are linear. However, when the temperature falls below 10 K, the dramatically increasing MR will severely interfere with the Hall signals, and the anomalous Hall effect becomes unobservable. To avoid the influence of MR, Hall resistivity curves and the related calculations of carrier concentration are all based on the measurements of the Hall effect above 10 K. Similar to the DMS Li(Zn,Mn)As, *p*-type conduction is expected, as Li^+ is doped to replaced Zn^{2+} to act as an acceptor. At 10 K, the carrier concentration is 3.5×10^{19} cm^{-3}. It becomes 5.09×10^{20}, 7.47×10^{20}, and 7.48×10^{20} cm^{-3} at 50 K, 70 K, and 100 K, respectively. The temperature-dependent carrier concentration shows common features with semiconductors, i.e., a monotonic increase in the carrier concentration at low temperatures and saturation at a relatively high temperature, which is consistent with the typical thermal excitation model. The carrier concentration is comparable to that of (Ga,Mn)As or $BaZn_2As_2$ (~10^{20} cm^{-3}). The exponential increase in hole density with increasing temperature is also consistent with the activation energy model behaviors of $\rho(T)$ curves.

4. Conclusions

In summary, we synthesized a new diluted magnetic semiconductor $Rb(Zn_{0.83}Li_{0.10}Mn_{0.07})_4As_3$ with a quasi-two-dimensional structure. With optimal charge and spin doping, it shows ferromagnetic transition with the highest Cuire temperature of 24 K. The analysis of critical behavior via Arrott plots and reliable self-consistent results confirm the long-range ferromagnetic ordering in $Rb(Zn_{1-x-y}Li_yMn_x)_4As_3$ single crystals. Owing to the low-dimensional structure, substantial magnetic anisotropy can be found when the field is parallel and perpendicular to the *c*-axis. Correspondingly, transverse magnetoresistance also exhibits large anisotropy. The most remarkable feature is that the $Rb(Zn_{0.83}Li_{0.10}Mn_{0.07})_4As_3$ single crystals show a positive MR of 85% at 0.4 T and a negative MR of −93% at 7 T. It is worth noting that these intriguing features are found at low temperatures or high fields. These rigorous conditions could be barriers to practical applications. Nevertheless, $Rb(Zn_{1-x-y}Li_yMn_x)_4As_3$ could stimulate future developments of analog DMS materials, but with a near-room-temperature Curie temperature, to benefit applications.

Author Contributions: Conceptualization, Y.P., Z.D. and C.J.; methodology, Y.P. and Z.D.; investigation, Y.P., L.S., G.Z., J.Z. (Jun Zhang), L.Z. (Jianfa Zhao), X.W. and Z.D.; writing—original draft preparation, Y.P. and Z.D.; writing—review and editing, Y.P., Z.D. and C.J.; supervision, Z.D. and C.J.;

project administration, Z.D. and C.J.; funding acquisition, Z.D. and C.J. All authors have read and agreed to the published version of the manuscript.

Funding: The work was supported by the National Key R&D Program of China (No. 2022YFA1403900), the NSF of China (No. 11974407 and 12104488), and the CAS Project for Young Scientists in Basic Research (No. YSBR-030).

Data Availability Statement: All data are available from the corresponding author upon reasonable request.

Conflicts of Interest: The authors declare no conflicts of interest.

References

1. Dietl, T. A ten-year perspective on dilute magnetic semiconductors and oxides. *Nat. Mater.* **2010**, *9*, 965. [CrossRef] [PubMed]
2. Furdyna, J.K. Diluted magnetic semiconductors. *J. Appl. Phys.* **1988**, *64*, R29. [CrossRef]
3. Hirohata, A.; Sukegawa, H.; Yanagihara, H.; Zutic, I.; Seki, T.; Mizukami, S.; Swaminathan, R. Roadmap for Emerging Materials for Spintronic Device Applications. *IEEE Trans. Magn.* **2015**, *51*, 0800511. [CrossRef]
4. Dietl, T.; Ohno, H. Dilute ferromagnetic semiconductors: Physics and spintronic structures. *Rev. Mod. Phys.* **2014**, *86*, 187–251. [CrossRef]
5. Zhao, X.; Dong, J.; Fu, L.; Gu, Y.; Zhang, R.; Yang, Q.; Xie, L.; Tang, Y.; Ning, F. $(Ba_{1-x}Na_x)F(Zn_{1-x}Mn_x)Sb$: A novel fluoride-antimonide magnetic semiconductor with decoupled charge and spin doping. *J. Semicond.* **2022**, *43*, 112501. [CrossRef]
6. Saito, H.; Yuasa, S.; Ando, K. Origin of the tunnel anisotropic magnetoresistance in $Ga_{1-x}Mn_xAs/ZnSe/Ga_{1-x}Mn_xAs$ magnetic tunnel junctions of II-VI/III-V heterostructures. *Phys. Rev. Lett.* **2005**, *95*, 086604. [CrossRef] [PubMed]
7. Liu, X.; Riney, L.; Guerra, J.; Powers, W.; Wang, J.; Furdyna, J.K.; Assaf, B.A. Colossal negative magnetoresistance from hopping in insulating ferromagnetic semiconductors. *J. Semicond.* **2022**, *43*, 112502. [CrossRef]
8. Žutić, I.; Zhou, T. Tailoring Magnetism in Semiconductors. *Sci. China-Phys. Mech. Astron.* **2018**, *61*, 067031. [CrossRef]
9. Chougule, B.K.; Begum, S. *Reference Module in Materials Science and Materials Engineering*; Elsevier: Amsterdam, The Netherlands, 2008; pp. 1–2.
10. Mathieu, R.; Akahoshi, D.; Asamitsu, A.; Tomioka, Y.; Tokura, Y. Colossal magnetoresistance without phase separation: Disorder-induced spin glass state and nanometer scale orbital-charge correlation in half doped manganites. *Phys. Rev. Lett.* **2004**, *93*, 227202. [CrossRef]
11. Ali, M.N.; Xiong, J.; Flynn, S.; Tao, J.; Gibson, Q.D.; Schoop, L.M.; Liang, T.; Haldolaarachchige, N.; Hirschberger, M.; Ong, N.P.; et al. Large, non-saturating magnetoresistance in WTe_2. *Nature* **2014**, *514*, 205–208. [CrossRef]
12. Csontos, M.; Wojtowicz, T.; Liu, X.; Dobrowolska, M.; Janko, B.; Furdyna, J.K.; Mihaly, G. Magnetic scattering of spin polarized carriers in (In, Mn)Sb dilute magnetic semiconductor. *Phys. Rev. Lett.* **2005**, *95*, 227203. [CrossRef]
13. Ohno, H. Properties of ferromagnetic III-V semiconductors. *J. Magn. Magn. Mater.* **1999**, *200*, 20. [CrossRef]
14. Glasbrenner, J.K.; Zutic, I.; Mazin, I.I. Theory of Mn-doped II-II-V semiconductors. *Phys. Rev. B* **2014**, *90*, 140403. [CrossRef]
15. Mašek, J.; Kudrnovský, J.; Máca, F.; Gallagher, B.; Campion, R.; Gregory, D.; Jungwirth, T. Dilute Moment n-Type Ferromagnetic Semiconductor Li(Zn,Mn)As. *Phys. Rev. Lett.* **2007**, *98*, 067202. [CrossRef] [PubMed]
16. Ohno, H. Making Nonmagnetic Semiconductors Ferromagnetic. *Science* **1998**, *281*, 6. [CrossRef] [PubMed]
17. Deng, Z.; Jin, C.Q.; Liu, Q.Q.; Wang, X.C.; Zhu, J.L.; Feng, S.M.; Chen, L.C.; Yu, R.C.; Arguello, C.; Goko, T.; et al. Li(Zn,Mn)As as a new generation ferromagnet based on a I-II-V semiconductor. *Nat. Commun.* **2011**, *2*, 422. [CrossRef] [PubMed]
18. Zhao, K.; Chen, B.; Zhao, G.; Yuan, Z.; Liu, Q.; Deng, Z.; Zhu, J.; Jin, C. Ferromagnetism at 230 K in $(Ba_{0.7}K_{0.3})(Zn_{0.85}Mn_{0.15})_2As_2$ diluted magnetic semiconductor. *Chin. Sci. Bull.* **2014**, *59*, 2524. [CrossRef]
19. Sun, F.; Zhao, G.Q.; Escanhoela, C.A.; Chen, B.J.; Kou, R.H.; Wang, Y.G.; Xiao, Y.M.; Chow, P.; Mao, H.K.; Haskel, D.; et al. Hole doping and pressure effects on the II-II-V-based diluted magnetic semiconductor $(Ba_{1-x}K_x)(Zn_{1-y}Mn_y)_2As_2$. *Phys. Rev. B* **2017**, *95*, 094412. [CrossRef]
20. Yang, X.; Li, Y.; Zhang, P.; Jiang, H.; Luo, Y.; Chen, Q.; Feng, C.; Cao, C.; Dai, J.; Tao, Q.; et al. K and Mn co-doped $BaCd_2As_2$: A hexagonal structured bulk diluted magnetic semiconductor with large magnetoresistance. *J. Appl. Phys.* **2013**, *114*, 223905. [CrossRef]
21. Yang, X.; Chen, Q.; Li, Y.; Wang, Z.; Bao, J.; Li, Y.; Tao, Q.; Cao, G.; Xu, Z.-A. $Sr_{0.9}K_{0.1}Zn_{1.8}Mn_{0.2}As_2$: A ferromagnetic semiconductor with colossal magnetoresistance. *Europhys. Lett.* **2014**, *107*, 67007. [CrossRef]
22. Hogl, P.; Matos-Abiague, A.; Zutic, I.; Fabian, J. Magnetoanisotropic Andreev reflection in ferromagnet-superconductor junctions. *Phys. Rev. Lett.* **2015**, *115*, 116601. [CrossRef]
23. Zutic, I.; Fabian, J.; Das Sarma, S. Spintronics: Fundamentals and applications. *Rev. Mod. Phys.* **2004**, *76*, 323. [CrossRef]
24. Zhao, K.; Deng, Z.; Wang, X.C.; Han, W.; Zhu, J.L.; Li, X.; Liu, Q.Q.; Yu, R.C.; Goko, T.; Frandsen, B.; et al. New diluted ferromagnetic semiconductor with Curie temperature up to 180 K and isostructural to the '122' iron-based superconductors. *Nat. Commun.* **2013**, *4*, 1442. [CrossRef] [PubMed]
25. He, H.; Tyson, C.; Bobev, S. Eight-Coordinated Arsenic in the Zintl Phases $RbCd_4As_3$ and $RbZn_4As_3$: Synthesis and Structural Characterization. *Inorg. Chem.* **2011**, *50*, 8375–8383. [CrossRef] [PubMed]

26. Hellmann, A.; Löhken, A.; Wurth, A.; Mewis, A. New Arsenides with $ThCr_2Si_2$-type or Related Structures: The Compounds ARh_2As_2 (A:Eu, Sr,Ba) and $BaZn_2As_2$. *Z. Naturforsch.* **2007**, *62*, 155. [CrossRef]
27. Toby, B.H. EXPGUI, a graphical user interface for GSAS. *J. Appl. Crystallogr.* **2001**, *34*, 210–213. [CrossRef]
28. Deng, Z.; Wang, X.; Wang, M.; Shen, F.; Zhang, J.; Chen, Y.; Feng, H.L.; Xu, J.; Peng, Y.; Li, W.; et al. Giant Exchange-Bias-Like Effect at Low Cooling Fields Induced by Pinned Magnetic Domains in Y_2NiIrO_6 Double Perovskite. *Adv. Mater.* **2023**, *35*, 2209759. [CrossRef]
29. Chen, L.; Yang, X.; Yang, F.; Zhao, J.; Misuraca, J.; Xiong, P.; von Molnar, S. Enhancing the Curie temperature of ferromagnetic semiconductor (Ga,Mn)As to 200 K via nanostructure engineering. *Nano Lett.* **2011**, *11*, 2584. [CrossRef]
30. Wei, Q.; Wang, H.; Zhao, X.; Zhao, J. Electron mobility anisotropy in (Al,Ga)Sb/InAs two-dimensional electron gases epitaxied on GaAs (001) substrates. *J. Semicond.* **2022**, *43*, 072101. [CrossRef]
31. Yu, S.; Liu, X.; Zhao, G.; Peng, Y.; Wang, X.; Zhao, J.; Li, W.; Deng, Z.; Furdyna, J.K.; Uemura, Y.J.; et al. Anomalous critical point behavior in dilute magnetic semiconductor $(Ca,Na)(Zn,Mn)_2Sb_2$. *Phys. Rev. Mater.* **2020**, *4*, 024411. [CrossRef]
32. Zhang, L.; Menzel, D.; Jin, C.; Du, H.; Ge, M.; Zhang, C.; Pi, L.; Tian, M.; Zhang, Y. Critical behavior of the single-crystal helimagnet MnSi. *Phys. Rev. B* **2015**, *91*, 024403. [CrossRef]
33. Kim, D.; Revaz, B.; Zink, B.L.; Hellman, F.; Rhyne, J.J.; Mitchell, J.F. Tricritical point and the doping dependence of the order of the ferromagnetic phase transition of $La_{1-x}Ca_xMnO_3$. *Phys. Rev. Lett.* **2002**, *89*, 227202. [CrossRef] [PubMed]
34. Campostrini, M.; Hasenbusch, M.; Pelissetto, A.; Rossi, P.; Vicari, E. Critical exponents and equation of state of the three-dimensional Heisenberg universality class. *Phys. Rev. B* **2002**, *65*, 144520. [CrossRef]
35. Wang, M.; Marshall, R.A.; Edmonds, K.W.; Rushforth, A.W.; Campion, R.P.; Gallagher, B.L. Three-dimensional Heisenberg critical behavior in the highly disordered dilute ferromagnetic semiconductor (Ga,Mn)As. *Phys. Rev. B* **2016**, *93*, 184417. [CrossRef]
36. Dietl, T.; Ohno, H.; Matsukura, F. Hole-mediated ferromagnetism in tetrahedrally coordinated semiconductors. *Phys. Rev. B* **2001**, *63*, 195205. [CrossRef]
37. Telford, E.J.; Dismukes, A.H.; Lee, K.; Cheng, M.; Wieteska, A.; Bartholomew, A.K.; Chen, Y.S.; Xu, X.; Pasupathy, A.N.; Zhu, X.; et al. Layered Antiferromagnetism Induces Large Negative Magnetoresistance in the van der Waals Semiconductor CrSBr. *Adv. Mater.* **2020**, *32*, e2003240. [CrossRef] [PubMed]
38. Ghimire, N.J.; Khan, M.A.; Botana, A.S.; Jiang, J.S.; Mitchell, J.F. Anisotropic angular magnetoresistance and Fermi surface topology of the candidate novel topological metal Pd_3Pb. *Phys. Rev. Mater.* **2018**, *2*, 081201. [CrossRef]
39. Wei, D. The room temperature ferromagnetism in highly strained two-dimensional magnetic semiconductors. *J. Semicond.* **2023**, *44*, 040401. [CrossRef]
40. Ohno, Y.; Young, D.K.; Beschoten, B.; Matsukura, F.; Ohno, H.; Awschalom, D.D. Electrical spin injection in a ferromagnetic semiconductor heterostructure. *Nature* **1999**, *402*, 790–792. [CrossRef]

nanomaterials

Article

The Influence of Capping Layers on Tunneling Magnetoresistance and Microstructure in CoFeB/MgO/CoFeB Magnetic Tunnel Junctions upon Annealing

Geunwoo Kim [1], Soogil Lee [1,2,*], Sanghwa Lee [1], Byonggwon Song [3], Byung-Kyu Lee [3], Duhyun Lee [3], Jin Seo Lee [4], Min Hyeok Lee [5], Young Keun Kim [5] and Byong-Guk Park [1,*]

1 Department of Materials Science and Engineering, Korea Advanced Institute of Science and Technology, Daejeon 34141, Republic of Korea
2 Department of Electronic Engineering, Gachon University, Seongnam 13120, Republic of Korea
3 Samsung Advanced Institute of Technology, Samsung Electronics, Suwon 16678, Republic of Korea
4 Department of Semiconductor Systems Engineering, Korea University, Seoul 02481, Republic of Korea
5 Department of Materials Science and Engineering, Korea University, Seoul 02481, Republic of Korea
* Correspondence: leesoogil83@gachon.ac.kr (S.L.); bgpark@kaist.ac.kr (B.-G.P.)

Abstract: This study investigates the effects of annealing on the tunnel magnetoresistance (TMR) ratio in CoFeB/MgO/CoFeB-based magnetic tunnel junctions (MTJs) with different capping layers and correlates them with microstructural changes. It is found that the capping layer plays an important role in determining the maximum TMR ratio and the corresponding annealing temperature (T_{ann}). For a Pt capping layer, the TMR reaches ~95% at a T_{ann} of 350 °C, then decreases upon a further increase in T_{ann}. A microstructural analysis reveals that the low TMR is due to severe intermixing in the Pt/CoFeB layers. On the other hand, when introducing a Ta capping layer with suppressed diffusion into the CoFeB layer, the TMR continues to increase with T_{ann} up to 400 °C, reaching ~250%. Our findings indicate that the proper selection of a capping layer can increase the annealing temperature of MTJs so that it becomes compatible with the complementary metal-oxide-semiconductor backend process.

Keywords: capping layer; magnetic tunnel junction; tunneling magnetoresistance; diffusion

Citation: Kim, G.; Lee, S.; Lee, S.; Song, B.; Lee, B.-K.; Lee, D.; Lee, J.S.; Lee, M.H.; Kim, Y.K.; Park, B.-G. The Influence of Capping Layers on Tunneling Magnetoresistance and Microstructure in CoFeB/MgO/CoFeB Magnetic Tunnel Junctions upon Annealing. *Nanomaterials* **2023**, *13*, 2591. https://doi.org/10.3390/nano13182591

Academic Editor: Julian Maria Gonzalez Estevezame

Received: 18 July 2023
Revised: 13 September 2023
Accepted: 16 September 2023
Published: 19 September 2023

1. Introduction

Magnetic tunnel junctions (MTJs) consisting of ferromagnet (FM)/tunnel barrier/FM structures have been studied extensively because they serve as a key element in various spintronic devices, including the read heads of hard disk drives and magnetic random-access memory devices (MRAMs) [1–5]. In MTJs, tunneling electrons are spin-polarized along the magnetization direction so that the tunneling probability increases (decreases) when the magnetizations of the two FMs are aligned in a parallel (antiparallel) configuration. This spin-dependent tunneling can be enhanced by introducing a (001)-oriented crystalline MgO barrier, where the Δ_1 Bloch state allows coherent tunneling [6–8]. A large tunneling magnetoresistance (TMR) ratio exceeding hundreds of percent as theoretically predicted in crystalline MgO-based MTJs has been experimentally demonstrated [9–12].

CoFeB is a widely employed FM electrode in sputter-grown MTJs [11–20] because, first, amorphous CoFeB favors the growth of the MgO close-packed (001) plane on top [12,15], and second, post-annealing causes boron to diffuse out of CoFeB and subsequently crystallize the amorphous CoFeB into CoFe on the MgO (001) texture [19,20]. It is also known that post-annealing at moderate temperatures improves the material and device properties of the MTJs by reducing defects in the MgO tunnel barrier or ferromagnet/MgO interfaces and/or by enhancing the crystallinity of the MgO layer. This leads to an increase in the TMR ratio or the perpendicular magnetic anisotropy (PMA). Therefore, in order to obtain a large TMR ratio in CoFeB/MgO/CoFeB-structured MTJs, it is essential to perform post-annealing at an elevated temperature. However, the annealing temperature

(T_{ann}) must not be excessive such that it causes atomic intermixing between the layers, which diminishes the TMR. Furthermore, for MTJs to be used in practical MRAM applications, they must maintain their characteristics at T_{ann} values above 400 °C, the range in which complementary metal-oxide-semiconductor (CMOS) backend integration takes place [21–23].

Many studies have examined annealing effects in antiferromagnet/CoFeB/MgO/CoFeB structures and the interfaces therein [9,12,16,19,24–41]. For example, the effect of the Ta capping layer has been extensively investigated. Upon annealing, the Ta capping layer effectively absorbs boron atoms from adjacent CoFeB layers, promoting the crystallization of the amorphous CoFeB layer. Consequently, MTJs with a Ta capping layer exhibit enhanced TMR ratios after annealing within a T_{ann} range of 300~500 °C [12,19,28–35]. The W or Hf capping layer demonstrates similar annealing effects in terms of the increased TMR because of the effective B absorption [36–38]. On the other hand, there have been few studies of the effects of annealing on Pt/CoFeB layers [39,40] or a Pt/CoFeB/MgO/CoFeB tunnel junction [41], where PMA deteriorates when samples are annealed with T_{ann} exceeding 300 °C. This was attributed to intermixing between the Pt and CoFeB layers. However, previous studies did not investigate the annealing effect on the TMR ratio in MTJs with a Pt capping layer, especially at the T_{ann} representative of the CMOS-compatible backend process.

Furthermore, heavy-metal (e.g., Pt, Ta, W)/ferromagnet bilayers have been extensively investigated because the spin currents generated in the heavy metal exert spin-orbit torque (SOT) on the ferromagnet and control its magnetization direction. SOT is being developed as a novel writing technology for energy-efficient MRAM [42–45]. Among such heavy elements, Pt is considered an excellent SOT material owing to its low resistivity and relatively large spin Hall angle, offering a distinct advantage in terms of power consumption for SOT-based spintronic devices over other spin-current source materials. Therefore, it is crucial to investigate the annealing effect on the TMR of CoFeB/MgO/CoFeB MTJs with a Pt capping layer compared to the effects on those with other capping layers.

In this study, we investigated the annealing-temperature dependence of TMR in IrMn/CoFeB/MgO/CoFeB MTJs with different heavy-metal capping layers of Pt, Ta, and W and correlated them with microstructural changes. For a Pt capping layer, the TMR reaches its maximum value of ~95% at T_{ann} of 350 °C and is then reduced as T_{ann} is increased further. Microstructural analyses reveal that annealing causes severe interdiffusion between the Pt and CoFeB layers, which is believed to be responsible for the reduced TMR value. Interestingly, when Ta is used as a capping layer, the TMR increases to ~250% and does not deteriorate even at a T_{ann} of 400 °C, meeting the thermal budget requirements of the CMOS backend process. MTJs with a W capping layer also exhibit similar behavior. Microstructural analyses confirm that intermixing is significantly suppressed in the CoFeB/Ta layer. Our study indicates that an appropriate capping layer can enhance the temperature dependence of the TMR in CoFeB/MgO/CoFeB MTJs.

2. Experimental Section

We fabricated MTJs consisting of a $Ni_{81}Fe_{19}$ (1 nm)/$Ir_{25}Mn_{75}$ (15 nm)/$Co_{32}Fe_{48}B_{20}$ (5 nm)/MgO (2 nm)/$Co_{32}Fe_{48}B_{20}$ (4 nm) structure with different capping layers of Pt, Ta, and W, as illustrated in Figure 1a. The films were deposited on Si/SiO_x (200 nm) substrates by magnetron sputtering at room temperature with a base pressure lower than 3.0×10^{-8} Torr. During the deposition, a magnetic field of 15 mT was applied to induce uniaxial anisotropy of the CoFeB layers. We deposited metal layers with a DC power of 30 W and a working pressure of 3 mTorr and a MgO layer with a RF power of 75 W and a working pressure of 10 mTorr. The aforementioned thickness of each layer is a nominal value calculated from the deposition rate of the layer, which was measured using a surface profiler (α-step) and an atomic force microscope. A thin NiFe layer was introduced underneath the IrMn layer to promote the exchange bias of the IrMn/CoFeB bilayer. After the film deposition process, we defined a dumbbell-shaped bottom electrode using photolithography and Ar ion milling techniques. The width of the bottom electrode is 10 μm. Then, pillar-shaped MTJs with a diameter of 5 μm were patterned by etching the top electrode and MgO tunnel barrier and in situ passivating a RF-deposited 70 nm thick SiO_x layer to electrically disconnect the bottom and top electrodes. Finally, the top electrode was formed by the deposition of a Cr (5 nm)/Au (100 nm) layer and subsequent lift-off process. Figure 1b presents a scanning electron microscope image of the MTJ device. After the fabrication of the device, MTJs were annealed at different temperatures (T_{ann}) ranging from 250 °C to 450 °C for 40 min in a vacuum condition. Here, the maximum T_{ann} of 450 °C was chosen by considering the thermal budget of the CMOS backend process. To establish exchange coupling at the IrMn/CoFeB interface, the annealing was conducted under a magnetic field of 100 mT, which is sufficient to saturate the magnetization of the CoFeB layer. The tunneling resistance was measured using a four-point geometry method with a constant reading current to apply a bias voltage of 10 mV at room temperature while sweeping the in-plane magnetic fields (B_{in}). The microstructures of the samples were analyzed through high-resolution scanning transmission electron microscopy (HR-STEM), energy dispersive X-ray spectroscopy (EDS), and X-ray diffraction (XRD).

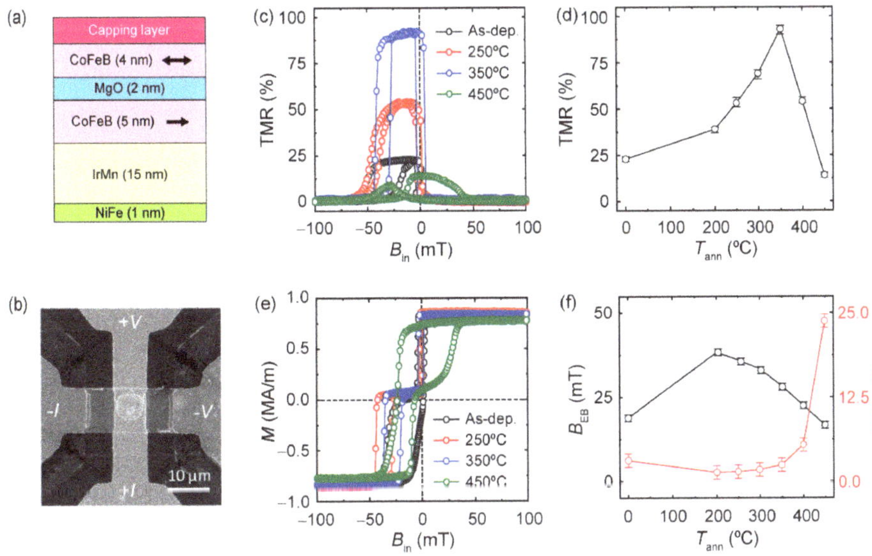

Figure 1. Annealing-temperature (T_{ann})-dependent TMR and magnetic properties in MTJs with a Pt capping layer. (**a**) Schematic of an MTJ with an IrMn/CoFeB/MgO/CoFeB structure. (**b**) Scanning electron microscope image of the MTJ device. (**c**) TMR curves of MTJs with different T_{ann}. (**d**) TMR as a function of T_{ann}. The error bars are obtained by averaging the measurement results from three MTJ devices. (**e**) Hysteresis loops of the MTJ films with different T_{ann}. (**f**) Exchange-bias field of the bottom CoFeB layer (B_{EB}) and coercivity of the top CoFeB layer (B_C^{top}) as a function of T_{ann}. The error bars represent the uncertainties caused by the resolution of the measurement systems.

3. Results and Discussion

First, we investigated the annealing effect on the TMR of an MTJ with a Pt (5 nm) capping layer. Figure 1c presents the TMR ratio versus the B_{in} curves measured for as-deposited and annealed samples at different T_{ann}. Here, the TMR ratio is defined as TMR (%) = [R_T (B_{in}) − R_T (B_{in} = 100 mT)]/[R_T (B_{in} = 100 mT)], where R_T is the tunnel resistance of the sample. Note that the magnetization directions of the two CoFeB layers are aligned in a parallel manner when B_{in} = 100 mT. The TMR curves show a clear distinction between low and high resistance states for the samples for which T_{ann} is below 350 °C, corresponding to the parallel and antiparallel alignment of the top and bottom magnetizations. However, it is severely modified for the sample with T_{ann} equal to 450 °C. Figure 1d shows the TMR as a function of T_{ann} when it ranges from 200 °C to 450 °C. This finding demonstrates two notable points. First, the TMR ratio of the MTJ with the Pt capping layer can only reach ~95% at T_{ann} = 350 °C, much lower than that of typical MgO-based MTJs [12]. Second, the TMR decreases drastically when T_{ann} exceeds 350 °C, which may be related to the modified TMR curve.

We also measured the magnetization curves of MTJs annealed at different T_{ann} by means of vibrating sample magnetometry. Figure 1e shows two hysteresis curves, one centered at $B_{in} = 0$ and the other centered at $B_{in} < 0$. The former (latter) corresponds to the magnetization of the top magnetically free (bottom exchange-biased) CoFeB layer. Notably, the hysteresis curve of the MTJ with T_{ann} = 450 °C is quite different from those of the other samples, a finding consistent with the TMR curve. Figure 1f shows the extracted exchange-bias field (B_{EB}) of the bottom CoFeB layer and coercivity (B_C^{top}) of the top CoFeB layer as a function of T_{ann}. Here, B_{EB} initially increases upon annealing at a T_{ann} of 200 °C, after which it gradually decreases upon a further increase in T_{ann}. We can understand the decrease in B_{EB} at a higher T_{ann} in terms of the diffusion of Mn from the antiferromagnet IrMn layer [25–27]. The behavior of B_{EB} with T_{ann} is not similar to that of the TMR, and the magnitude of B_{EB} still shows a finite value even after annealing at a T_{ann} of 450 °C, suggesting that the decrease in B_{EB} is not the main cause of the reduction in the TMR ratio at a high T_{ann}. On the other hand, the B_C^{top} shows different behavior; the values remain unchanged for both the as-deposited and the annealed samples with T_{ann} values up to 350 °C. However, it increases drastically upon annealing when T_{ann} exceeds 400 °C, with the magnitude becoming ~24 mT, which is more than 10 times greater than those of the samples with lower T_{ann}s. It was also found that the magnetization of the top CoFeB layer decreases slightly after annealing at a T_{ann} of 450 °C. This, together with the large increase in B_C^{top}, which causes the antiparallel magnetic alignment to be less pronounced, may be responsible for the reduced TMR ratio.

To understand the T_{ann} dependence of the magnetic properties of the top CoFeB layer and the associated TMR ratio, we conducted microstructural analyses using cross-sectional STEM and EDS. Figure 2a,b correspondingly show bright-field STEM images of the as-deposited sample and its annealed counterpart when T_{ann} = 450 °C. Both images show clear interfaces of the IrMn/bottom CoFeB/MgO/top CoFeB layers. However, we cannot clearly distinguish the interface between the top CoFeB and Pt layers in either film. To clarify the interface quality, we also investigated high-angle annular dark-field (HAADF) images, in which the contrast is correlated with the atomic number. Figure 2c,d present HAADF-STEM images of the as-deposited and annealed films, respectively. The as-deposited film exhibits a contrast difference, albeit weak, between the top CoFeB and Pt layers. However, the annealed film at a T_{ann} of 450 °C shows no contrast difference between the layers. This indicates significant atomic intermixing between the CoFeB/Pt layers.

To assess this intermixing, we construct element mapping images by EDS. Figure 2e,f show images of the as-deposited and annealed films, respectively. Here, yellow, green, and blue represent Pt, Co, and Fe atoms, respectively. It is observed that yellow appears within the top CoFeB layer, even in the as-deposited sample (Figure 2e), indicating that a nonnegligible amount of Pt diffused into the top CoFeB layer during the deposition of the films. The yellow contrast in the top CoFeB layer becomes significantly stronger in Figure 2f, showing that more Pt atoms diffused into the top CoFeB layer by annealing. Note that the contrast of the Co and Fe atoms is not noticeably changed by annealing. We consider Pt diffusion to be the main cause of the reduced TMR. The alloy formation of the diffused Pt with CoFeB can explain the annealing-induced changes in the magnetic properties: an increase in B_C^{top} and a decrease in the magnetization of the top CoFeB layer. In addition to increasing B_C^{top}, which makes it difficult to distinguish between parallel and antiparallel magnetization states, Pt diffusion may cause other mechanisms that, in turn, decrease the TMR. One may modify the band structure of CoFeB by adding nonmagnetic Pt, reducing the spin polarization value of the CoFeB layer [46]. Another is that Pt with a face-centered cubic (fcc) structure may prevent CoFeB from being epitaxially matched with the body-centered cubic (bcc) MgO (001) structure, reducing the degree of coherent tunneling [39]. Note that the diffusion of Pt into CoFeB must be suppressed in order to utilize Pt as a spin current source in SOT-based spintronic devices. The interdiffusion within the Pt/CoFeB layers can be mitigated by employing rapid thermal annealing (RTA), which reduces the annealing time [47], and by introducing an insertion layer between the Pt and CoFeB layers. The insertion layer must serve as a diffusion barrier while also enabling the transfer of the spin current generated in the Pt layer to the CoFeB layer without a significant loss [28,31,48].

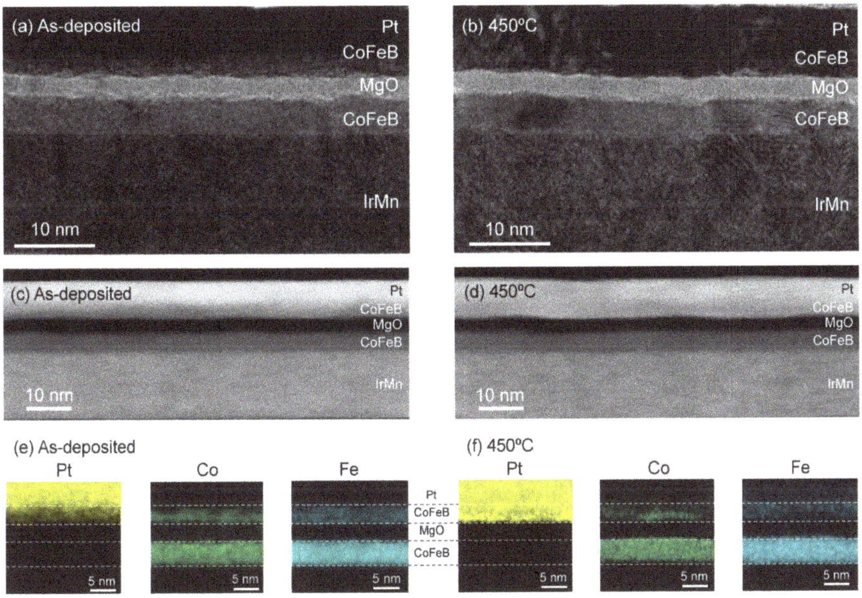

Figure 2. Microstructural analysis using cross-sectional STEM and EDS. (**a**,**b**) Bright-field STEM images of a MTJ with a Pt capping layer; as-deposited (**a**) and annealed films (**b**). (**c**,**d**) HAADF-STEM images of as-deposited (**c**) and annealed (**d**) films. (**e**,**f**) EDS element mapping images of Pt, Co, and Fe atoms; as-deposited (**e**) and annealed films (**f**). T_{ann} = 450 °C.

Next, we examine other capping layers of Ta or W that were introduced between the CoFeB and Pt layers while keeping the remaining layers the same. The layer structure is $Ni_{81}Fe_{19}$ (1 nm)/$Ir_{25}Mn_{75}$ (15 nm)/$Co_{32}Fe_{48}B_{20}$ (5 nm)/MgO (2 nm)/$Co_{32}Fe_{48}B_{20}$ (4 nm)/W or Ta (5 nm)/Pt (5 nm). Hereafter, we refer to the MTJs with a Ta (W) capping layer as Ta-(W-)-capped MTJs. Figure 3a,b show the TMR curves measured at T_{ann} of 400 °C and 450 °C for the Ta- and W-capped MTJs, respectively. Note that Ta and W are extensively utilized as spin-current sources, akin to Pt in SOT-based devices [42,43,49,50]. Unlike the MTJ with a Pt capping layer, the TMR curves of these samples retain their shape with well-defined parallel and antiparallel magnetization states, even after annealing at 450 °C. Figure 3c presents the TMR value as a function of T_{ann} when it ranges from 200 °C to 450 °C, demonstrating a significant improvement in the TMR ratio and the corresponding T_{ann} dependence compared to the Pt-capped MTJs. The maximum TMR ratios become 250% and 190% for the Ta- and W-capped MTJs, respectively. Moreover, T_{ann} leading to the maximum TMR increases to 400 °C for both samples. The slight decrease in the TMR ratio at T_{ann} = 450 °C may be due to Mn diffusion from the bottom IrMn layer at a high T_{ann} [25–27]. Figure 3d,e show the magnetization curves of the Ta- and W-capped MTJs films, respectively, with different T_{ann}s of 400 °C and 450 °C. Unlike the sample with a Pt capping layer (Figure 1d), these samples exhibit two clearly distinguished hysteresis loops for all T_{ann}s. Figure 3f shows the extracted B_{EB} of the bottom CoFeB and B_C^{top} as a function of T_{ann}. The B_{EB} behavior with regard to the T_{ann} of the samples with the Ta and W capping layers is very similar to that of the Pt sample; B_{EB} initially increases upon annealing at a T_{ann} of 200 °C, followed by a gradual decrease with a further increase in T_{ann}. This result confirms that the change in B_{EB} with T_{ann} is not the main cause of the T_{ann}-dependent TMR. On the other hand, the B_C^{top} values do not significantly change for the two samples over the entire T_{ann} range up to 450 °C. This is in stark contrast to the Pt sample (Figure 1e). In addition, the magnetizations of the top CoFeB layer of the W- and Ta-capped samples remain unchanged after annealing at a T_{ann} of 450 °C. These results demonstrate that the magnetic properties of the top CoFeB with the Ta or W capping layer are robust to the annealing at a T_{ann} of up to 450 °C, possibly because of the suppressed atomic diffusion. This is in line with the enhancement of the T_{ann} dependence of the TMR.

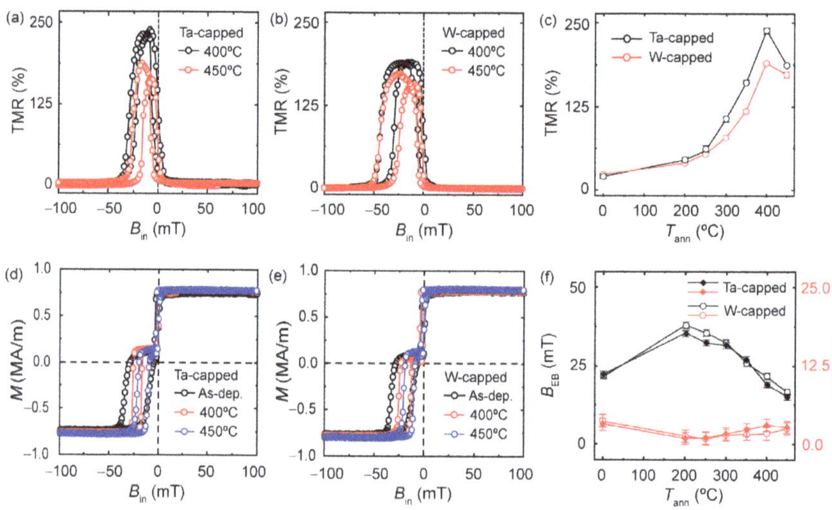

Figure 3. Annealing-temperature (T_{ann})-dependent TMR and magnetic properties in MTJs with Ta and W capping layers. The layer structure is $Ni_{81}Fe_{19}$ (1 nm)/$Ir_{25}Mn_{75}$ (15 nm)/$Co_{32}Fe_{48}B_{20}$ (5 nm)/MgO (2 nm)/$Co_{32}Fe_{48}B_{20}$ (4 nm)/W or Ta (4 nm)/Pt (5 nm). (**a**,**b**) TMR curves of MTJs with a Ta capping (**a**) and a W capping (**b**) layer annealed at a T_{ann} of 400 °C and 450 °C, respectively. (**c**) TMR as a function of T_{ann} in Ta- and W-capped MTJs. The error bars are obtained by averaging the measurement results from three MTJ devices. (**d**,**e**) Hysteresis loops of the Ta-capped (**d**) and W-capped (**e**) MTJ films with different T_{ann}. (**f**) Exchange-bias field of the bottom CoFeB layer (B_{EB}) and coercivity of the top CoFeB layer (B_C^{top}) as a function of T_{ann} for the Ta- and W-capped MTJ films. The error bars represent the uncertainties caused by the resolution of the measurement systems.

Additionally, we investigated the microstructures of Ta-capped MTJ films upon annealing using cross-sectional STEM and EDS. Figure 4a,b show bright-field STEM images of as-deposited and annealed (T_{ann} = 450 °C) films, respectively. Both images reveal well-defined layers and sharp interfaces, including the top CoFeB/Ta layers. The HAADF-STEM images as shown in Figure 4c,d exhibit a clear contrast difference for all layers, confirming that the top CoFeB/Ta layer structure is maintained after annealing at T_{ann} = 450 °C. Figure 4e,f show EDS element mapping images of the as-deposited and annealed films, respectively. Here, red, green, and blue represent Ta, Co, and Fe atoms, respectively. No red color is observed on the top CoFeB layer, indicating that the diffusion of Ta atoms into the top CoFeB layer after annealing at 450 °C was suppressed. This result is in stark contrast to that of the Pt-capped MTJ shown in Figure 2. On the other hand, the green and blue contrasts corresponding to Co and Fe atoms did not change significantly after annealing, similar to the Pt-capped MTJs. The W capping layer shows diffusion behavior similar to that of the Ta layer (Supplementary Information S1). Here, the Pt capping layer more easily intermixes with CoFeB than the Ta or W capping layer, possibly because of the greater solubility of Pt in Co or Fe compared to W or Ta in Co or Fe [51–56]. Furthermore, X-ray photoemission spectroscopy measurements reveal that the Ta layer acts as an effective boron absorber during the annealing process, thereby facilitating the crystallization of the CoFeB layer (Supplementary Information S2). These results indicate that the Ta capping layer effectively serves as an appropriate capping layer that suppresses atomic diffusion during the annealing process, consequently enhancing the T_{ann} dependence of the TMR and satisfying the thermal budget requirement of the CMOS backend process.

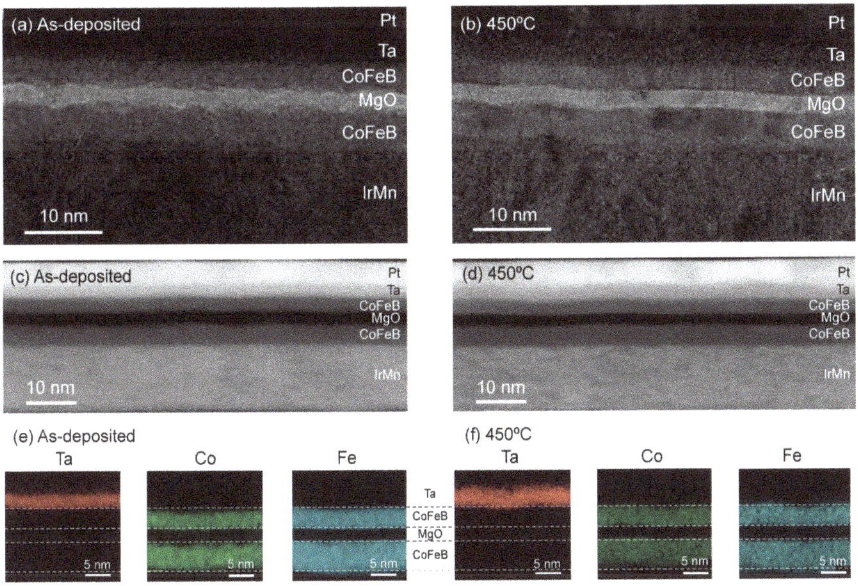

Figure 4. Bright-field STEM images of the Ta-capped MTJ structure of as-deposited (**a**) and annealed (**b**) films. The layer structure is $Ni_{81}Fe_{19}$ (1 nm)/$Ir_{25}Mn_{75}$ (15 nm)/$Co_{32}Fe_{48}B_{20}$ (5 nm)/MgO (2 nm)/$Co_{32}Fe_{48}B_{20}$ (4 nm)/Ta (4 nm)/Pt (5 nm). (**c**,**d**) HAADF-STEM images of the Ta-capped MTJ structure of as-deposited (**c**) and annealed (**d**) films. (**e**,**f**) EDS element mapping images of as-deposited (**e**) and annealed (**f**) films when T_{ann} = 450 °C.

We also examined microstructural changes upon annealing using the high-resolution θ–2θ XRD measurements. Note that we used thicker MgO (10 nm) and top CoFeB (10 nm) layers than those in the MTJ devices to enhance the XRD signal. Figure 5a shows the XRD spectra of the Pt samples with different T_{ann}s of 350 and 450 °C. It was found that the MgO (200) peak appears after annealing with T_{ann} = 350 °C and that the corresponding intensity decreases at T_{ann} = 450 °C, while the FePt (111) peak appears with 2θ = 44.21° [57]. This result provides the microstructural origins of the significant reduction in the TMR ratio at a high T_{ann}: the degradation of the MgO (001) crystal structure and the formation of a second phase of FePt caused by the diffusion of Pt atoms. On the other hand, Figure 5b,c show the XRD measurement results of the Ta and W samples, where the MgO (200) peaks remain unchanged and no additional peak emerges after annealing at T_{ann} = 450 °C. This again confirms that interdiffusion in the Ta and W samples is significantly suppressed even at a T_{ann} of 450 °C. These results are consistent with the enhanced temperature dependence of the TMR ratio of the Ta- and W-capped MTJs.

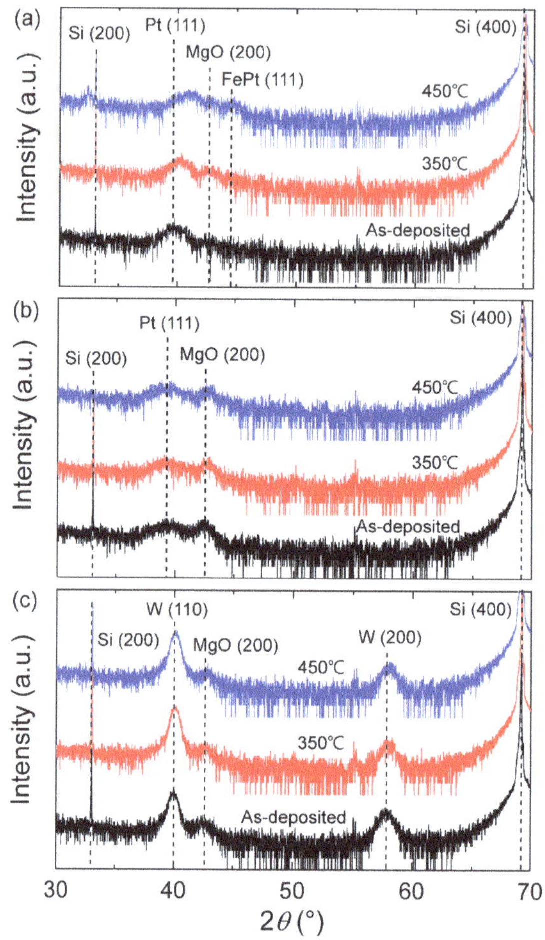

Figure 5. (**a**–**c**) The X-ray diffraction spectra of the samples with different capping layers. The film consists of a Ta (2 nm)/CoFeB (3 nm)/MgO (10 nm)/CoFeB (10 nm) structure with Pt (5 nm) (**a**), Ta (4 nm)/Pt (3 nm) (**b**), and W(4 nm)/Pt (3 nm) (**c**). T_{ann} is 350 °C and 450 °C.

4. Conclusions

In this study, we investigated the annealing effect on TMR in IrMn/CoFeB/MgO/CoFeB MTJs with different heavy-metal capping layers. For an MTJ with a Pt capping layer, the TMR reaches a maximum value of ~95% with T_{ann} = 350 °C, decreasing drastically when T_{ann} is increased further. With microstructural analyses using STEM and EDS, we attributed this low TMR to significant intermixing between the Pt and top CoFeB layers. This suggests that in order for Pt to be used as a SOT material, the property degradation after annealing at elevated temperatures must be overcome. Unlike the Pt capping layer, when introducing Ta and W capping layers with suppressed diffusion into the top CoFeB, the TMR ratio and corresponding temperature dependence significantly improved, showing a maximum TMR of ~250% at T_{ann} = 400 °C. Our study highlights the importance of the capping layer, which can significantly affect the MTJ device performance capabilities, suggesting that a proper capping layer will enhance the temperature dependence of the TMR so that applicability to the CMOS backend process can be realized.

Supplementary Materials: The following supporting information can be downloaded at https://www.mdpi.com/article/10.3390/nano13182591/s1. Supplementary Information S1. The diffusion of the W capping layer. Supplementary Information S2. The diffusion of B atoms and crystallization of the CoFeB layers

Author Contributions: Conceptualization, G.K., S.L. (Soogil Lee) and B.-G.P.; investigation, G.K., S.L. (Sanghwa Lee), B.S., D.L. and B.-K.L.; resources, G.K., J.S.L., M.H.L. and Y.K.K.; supervision, S.L. (Soogil Lee) and B.-G.P.; writing—original draft, G.K.; writing—review and editing, G.K, S.L. (Soogil Lee), Y.K.K. and B.-G.P. All authors have read and agreed to the published version of the manuscript.

Funding: This research was funded by the National Research Foundation of Korea (NRF-2022M3H4A1A04096339 and RS-2023-00261042).

Data Availability Statement: Data are available from the correspondence author.

Conflicts of Interest: The authors declare no conflict of interest.

References

1. Zhu, J.-G.; Park, C. Magnetic tunnel junctions. *Mater Today* **2006**, *9*, 36–45. [CrossRef]
2. Yuasa, S.; Fukushima, A.; Yakushiji, K.; Nozaki, T.; Konoto, M.; Maehara, H.; Kubota, H.; Taniguchi, T.; Arai, H.; Imamura, H.; et al. Future prospects of MRAM technologies. In Proceedings of the 2013 IEEE International Electron Devices Meeting, Washington, DC, USA, 9–11 December 2013; pp. 3.1.1–3.1.4. [CrossRef]
3. Hirohata, A.; Yamada, K.; Nakatani, Y.; Prejbeanu, I.-L.; Diény, B.; Pirro, P.; Hillebrands, B. Review on spintronics: Principles and device applications. *J. Magn. Magn. Mater.* **2020**, *509*, 166711. [CrossRef]
4. Dieny, B.; Prejbeanu, I.L.; Garello, K.; Gambardella, P.; Freitas, P.; Lehndorff, R.; Raberg, W.; Ebels, U.; Demokritov, S.O.; Akerman, J.; et al. Opportunities and challenges for spintronics in the microelectronics industry. *Nat. Electron.* **2020**, *3*, 446–459. [CrossRef]
5. Yuasa, S.; Djayaprawira, D.D. Giant tunnel magnetoresistance in magnetic tunnel junctions with a crystalline MgO(001) barrier. *J. Phys. D Appl. Phys.* **2007**, *40*, R337–R354. [CrossRef]
6. Butler, W.H.; Zhang, X.-G.; Schulthess, T.C.; MacLaren, J.M. Spin-dependent tunneling conductance of Fe/MgO/Fe sandwiches. *Phys. Rev. B* **2001**, *63*, 054416. [CrossRef]
7. Mathon, J.; Umerski, A. Theory of tunneling magnetoresistance of an epitaxial Fe/MgO/Fe(001) junction. *Phys. Rev. B.* **2001**, *63*, 200403. [CrossRef]
8. Zhang, X.-G.; Butler, W.H. Large magnetoresistance in bcc Co/MgO/Co and FeCo/MgO/FeCo tunnel junctions. *Phys. Rev. B* **2004**, *70*, 172407. [CrossRef]
9. Parkin, S.S.P.; Kaiser, C.; Panchula, A.; Rice, P.M.; Hughes, B.; Samant, M.; Yang, S.-H. Giant tunnelling magnetoresistance at room temperature with MgO (100) tunnel barriers. *Nat. Mater.* **2004**, *3*, 862–867. [CrossRef]
10. Yuasa, S.; Nagahama, T.; Fukushima, A.; Suzuki, Y.; Ando, K. Giant room-temperature magnetoresistance in single-crystal Fe/MgO/Fe magnetic tunnel junctions. *Nat. Mater.* **2004**, *3*, 868–871. [CrossRef]
11. Hayakawa, J.; Ikeda, S.; Matsukura, F.; Takahashi, H.; Ohno, H. Dependence of Giant Tunnel Magnetoresistance of Sputtered CoFeB/MgO/CoFeB Magnetic Tunnel Junctions on MgO Barrier Thickness and Annealing Temperature. *Jpn. J. Appl. Phys.* **2005**, *4*, L587–L589. [CrossRef]
12. Djayaprawira, D.D.; Tsunekawa, K.; Nagai, M.; Maehara, H.; Yamagata, S.; Watanabe, N.; Yuasa, S.; Suzuki, Y.; Ando, K. 230% room-temperature magnetoresistance in CoFeB/MgO/CoFeB magnetic tunnel junctions. *Appl. Phys. Lett.* **2005**, *86*, 092502. [CrossRef]
13. Yuasa, S.; Suzuki, Y.; Katayama, T.; Ando, K. Characterization of growth and crystallization processes in CoFeB/MgO/CoFeB magnetic tunnel junction structure by reflective high-energy electron diffraction. *Appl. Phys. Lett.* **2005**, *87*, 242503. [CrossRef]
14. Ikeda, S.; Hayakawa, J.; Lee, Y.M.; Tanikawa, T.; Matsukura, F.; Ohno, H. Tunnel magnetoresistance in MgO-barrier tunnel junctions with bcc-CoFe(B) and fcc-CoFe free layers. *J. Appl. Phys.* **2006**, *99*, 08A907. [CrossRef]
15. Bae, J.Y.; Lim, W.C.; Kim, H.J.; Lee, T.D.; Kim, K.W.; Kim, T.W. Compositional change of MgO barrier and interface in CoFeB/MgO/CoFeB tunnel junction after annealing. *J. Appl. Phys.* **2006**, *99*, 08T316. [CrossRef]
16. Read, J.C.; Mather, P.G.; Buhrman, R.A. X-ray photoemission study of CoFeB/MgO thin film bilayers. *Appl. Phys. Lett.* **2007**, *90*, 132503. [CrossRef]
17. Mizuguchi, M.; Suzuki, Y.; Nagahama, T.; Yuasa, S. In situ scanning tunneling microscopy observations of polycrystalline MgO(001) tunneling barriers grown on amorphous CoFeB electrode. *Appl. Phys. Lett.* **2007**, *90*, 012507. [CrossRef]
18. Cha, J.J.; Read, J.C.; Buhrman, R.A.; Muller, D.A. Spatially resolved electron energy-loss spectroscopy of electron-beam grown and sputtered CoFeB/MgO/CoFeB magnetic tunnel junctions. *Appl. Phys. Lett.* **2007**, *90*, 062516. [CrossRef]
19. Ikeda, S.; Hayakawa, J.; Ashizawa, Y.; Lee, Y.M.; Miura, K.; Hasegawa, H.; Tsunoda, M.; Matsukura, F.; Ohno, H. Tunnel magnetoresistance of 604% at 300K by suppression of Ta diffusion in CoFeB/MgO/CoFeB pseudo-spin-valves annealed at high temperature. *Appl. Phys. Lett.* **2008**, *93*, 082508. [CrossRef]

20. Ikeda, S.; Miura, K.; Yamamoto, H.; Mizunuma, K.; Gan, H.D.; Endo, M.; Kanai, S.; Hayakawa, J.; Matsukura, F.; Ohno, H. A perpendicular-anisotropy CoFeB–MgO magnetic tunnel junction. *Nat. Mater.* **2010**, *9*, 721–724. [CrossRef]
21. Cardoso, S.; Freitas, P.P.; de Jesus, C.; Soares, J.C. High thermal stability tunnel junctions. *J. Appl. Phys.* **2000**, *87*, 6058–6060. [CrossRef]
22. Yamane, K.; Higo, Y.; Uchida, H.; Nanba, Y.; Sasaki, S.; Ohmori, H.; Bessho, K.; Hosomi, M. Spin Torque Switching of Perpendicularly Magnetized CoFeB-Based Tunnel Junctions With High Thermal Tolerance. *IEEE Trans. Magn.* **2013**, *49*, 4335–4338. [CrossRef]
23. Thomas, L.; Jan, G.; Zhu, J.; Liu, H.; Lee, Y.-J.; Le, S.; Tong, R.-Y.; Pi, K.; Wang, Y.-J.; Shen, D.; et al. Perpendicular spin transfer torque magnetic random access memories with high spin torque efficiency and thermal stability for embedded applications. *J. Appl. Phys.* **2014**, *115*, 172615. [CrossRef]
24. You, C.Y.; Ohkubo, T.; Takahashi, Y.K.; Hono, K. Boron segregation in crystallized MgO/amorphous-$Co_{40}Fe_{40}B_{20}$ thin films. *J. Appl. Phys.* **2008**, *104*, 033517. [CrossRef]
25. Lee, J.H.; Kim, S.J.; Yoon, C.S.; Kim, C.K.; Park, B.G.; Lee, T.D. Thermal stability of the exchanged biased CoFe/IrMn electrode for the magnetic tunnel junction as a function of CoFe thickness. *J. Appl. Phys.* **2002**, *92*, 6241–6244. [CrossRef]
26. Lee, Y.M.; Hayakawa, J.; Ikeda, S.; Matsukura, F.; Ohno, H. Giant tunnel magnetoresistance and high annealing stability in CoFeB/MgO/CoFeB magnetic tunnel junctions with synthetic pinned layer. *Appl. Phys. Lett.* **2006**, *89*, 042506. [CrossRef]
27. Wang, Y.; Zeng, Z.M.; Han, X.F.; Zhang, X.G.; Sun, X.C.; Zhang, Z. Temperature-dependent Mn-diffusion modes in CoFeB- and CoFe-based magnetic tunnel junctions: Electron-microscopy studies. *Phys. Rev. B* **2007**, *75*, 214424. [CrossRef]
28. Miyajima, T.; Ibusuki, T.; Umehara, S.; Sato, M.; Eguchi, S.; Tsukada, M.; Kataoka, Y. Transmission electron microscopy study on the crystallization and boron distribution of CoFeB/MgO/CoFeB magnetic tunnel junctions with various capping layers. *Appl. Phys. Lett.* **2009**, *94*, 122501. [CrossRef]
29. Pong, P.W.T.; Egelhoff, W.F. Enhancement of tunneling magnetoresistance by optimization of capping layer thicknesses in CoFeB/MgO/CoFeB magnetic tunnel junctions. *J. Appl. Phys.* **2009**, *105*, 07C915. [CrossRef]
30. Mizunuma, K.; Ikeda, S.; Sato, H.; Yamanouchi, M.; Gan, H.; Miura, K.; Yamamoto, H.; Hayakawa, J.; Matsukura, F.; Ohno, H. Tunnel magnetoresistance properties and annealing stability in perpendicular anisotropy MgO-based magnetic tunnel junctions with different stack structures. *J. Appl. Phys.* **2011**, *109*, 07C711. [CrossRef]
31. Karthik, S.V.; Takahashi, Y.K.; Ohkubo, T.; Hono, K.; Gan, H.D.; Ikeda, S.; Ohno, H. Transmission electron microscopy study on the effect of various capping layers on CoFeB/MgO/CoFeB pseudo spin valves annealed at different temperatures. *J. Appl. Phys.* **2012**, *111*, 083922. [CrossRef]
32. Bouchikhaoui, H.; Stender, P.; Akemeier, D.; Baither, D.; Hono, K.; Hütten, A.; Schmitz, G. On the role of Ta cap in the recrystallization process of CoFeB layers. *Appl. Phys. Lett.* **2013**, *103*, 142412. [CrossRef]
33. Almasi, H.; Hickey, D.R.; Newhouse-Illige, T.; Xu, M.; Rosales, M.R.; Nahar, S.; Held, J.T.; Mkhoyan, K.A.; Wang, W.G. Enhanced tunneling magnetoresistance and perpendicular magnetic anisotropy in Mo/CoFeB/MgO magnetic tunnel junctions. *Appl. Phys. Lett.* **2015**, *106*, 182406. [CrossRef]
34. Bouchikhaoui, H.; Stender, P.; Balogh, Z.; Baither, D.; Hütten, A.; Hono, K.; Schmitz, G. Nano-analysis of Ta/FeCoB/MgO tunnel magneto resistance structures. *Acta Mater.* **2016**, *116*, 298–307. [CrossRef]
35. Yamamoto, T.; Nozaki, T.; Yakushiji, K.; Tamaru, S.; Kubota, H.; Fukushima, A.; Yuasa, S. Perpendicular magnetic anisotropy and its voltage control in MgO/CoFeB/MgO junctions with atomically thin Ta adhesion layers. *Acta Mater.* **2021**, *216*, 117097. [CrossRef]
36. Chatterjee, J.; Sousa, R.C.; Perrissin, N.; Auffret, S.; Ducruet, C.; Dieny, B. Enhanced annealing stability and perpendicular magnetic anisotropy in perpendicular magnetic tunnel junctions using W layer. *Appl. Phys. Lett.* **2017**, *110*, 202401. [CrossRef]
37. Xu, X.; Mukaiyama, K.; Kasai, S.; Ohkubo, T.; Hono, K. Impact of boron diffusion at MgO grain boundaries on magneto-transport properties of MgO/CoFeB/W magnetic tunnel junctions. *Acta Mater.* **2018**, *161*, 360. [CrossRef]
38. Manos, O.; Böhnke, A.; Bougiatioti, P.; Klett, R.; Rott, K.; Niesen, A.; Schmalhorst, J.-M.; Reiss, G. Tunneling magnetoresistance of perpendicular CoFeB-based junctions with exchange bias. *J. Appl. Phys.* **2017**, *122*, 103904. [CrossRef]
39. Zhu, Y.; Zhang, Z.; Ma, B.; Jin, Q.Y. Thermal stability of CoFeB/Pt multilayers with perpendicular magnetic anisotropy. *J. Appl. Phys.* **2012**, *111*, 07C106. [CrossRef]
40. Cestarollo, L.; Srinivasan, K.; El-Ghazaly, A. Investigation of perpendicular magnetic anisotropy in Pt/Co20Fe60B20/Pt multilayer structures. *J. Magn. Magn. Mater.* **2022**, *562*, 169825. [CrossRef]
41. Lee, D.-Y.; Shim, T.-H.; Park, J.-G. Effects of Pt capping layer on perpendicular magnet anisotropy in pseudo-spin valves of Ta/CoFeB/MgO/CoFeB/Pt magnetic-tunneling junctions. *Appl. Phys. Lett.* **2013**, *102*, 212409. [CrossRef]
42. Miron, I.M.; Garello, K.; Gaudin, G.; Zermatten, P.-J.; Costache, M.V.; Auffret, S.; Bandiera, S.; Rodmacq, B.; Schuhl, A.; Gambardella, P. Perpendicular switching of a single ferromagnetic layer induced by in-plane current injection. *Nature* **2011**, *476*, 189–193. [CrossRef] [PubMed]
43. Liu, L.; Pai, C.-F.; Li, Y.; Tseng, H.W.; Ralph, D.C.; Buhrman, R.A. Spin-Torque Switching with the Giant Spin Hall Effect of Tantalum. *Science* **2012**, *336*, 555–558. [CrossRef] [PubMed]
44. Manchon, A.; Železný, J.; Miron, I.M.; Jungwirth, T.; Sinova, J.; Thiaville, A.; Garello, K.; Gambardella, P. Current-induced spin-orbit torques in ferromagnetic and antiferromagnetic systems. *Rev. Mod. Phys.* **2019**, *91*, 035004. [CrossRef]

45. Ryu, J.; Lee, S.; Lee, K.; Park, B. Current-Induced Spin–Orbit Torques for Spintronic Applications. *Adv. Mater.* **2020**, *32*, e1907148. [CrossRef] [PubMed]
46. Messina, L.; Nastar, M.; Sandberg, N.; Olsson, P. Systematic electronic-structure investigation of substitutional impurity diffusion and flux coupling in bcc iron. *Phys. Rev. B* **2016**, *93*, 184302. [CrossRef]
47. Almasi, H.; Sun, C.L.; Li, X.; Newhouse-Illige, T.; Bi, C.; Price, K.C.; Nahar, S.; Grezes, C.; Hu, Q.; Amiri, P.K.; et al. Perpendicular magnetic tunnel junction with W seed and capping layers. *J. Appl. Phys.* **2017**, *121*, 153902. [CrossRef]
48. Lee, H.-Y.; Kim, S.; Park, J.-Y.; Oh, Y.-W.; Park, S.-Y.; Ham, W.; Kotani, Y.; Nakamura, T.; Suzuki, M.; Ono, T.; et al. Enhanced spin–orbit torque via interface engineering in Pt/CoFeB/MgO heterostructures. *APL Mater.* **2019**, *7*, 031110. [CrossRef]
49. Cho, S.; Baek, S.-H.C.; Lee, K.-D.; Jo, Y.; Park, B.-G. Large spin Hall magnetoresistance and its correlation to the spin-orbit torque in W/CoFeB/MgO structures. *Sci. Rep.* **2015**, *5*, 14668. [CrossRef]
50. Cha, I.H.; Lee, M.H.; Kim, G.W.; Kim, T.; Kim, Y.K. Spin-orbit torque efficiency in Ta or W/Ta-W/CoFeB junctions. *Mater. Res. Express* **2021**, *8*, 106102. [CrossRef]
51. Okamoto, H. Co-W (Cobalt-Tungsten). *J. Phase Equilibria Diffus.* **2008**, *29*, 119. [CrossRef]
52. Okamoto, H. Fe-Ta (Iron-Tantalum). *J. Phase Equilibria Diffus.* **2013**, *34*, 165–166. [CrossRef]
53. Shinagawa, K.; Chinen, H.; Omori, T.; Oikawa, K.; Ohnuma, I.; Ishida, K.; Kainuma, R. Phase equilibria and thermodynamic calculation of the Co–Ta binary system. *Intermetallics* **2014**, *49*, 87–97. [CrossRef]
54. Okamoto, H. Supplemental Literature Review of Binary Phase Diagrams: B-Fe, Cr-Zr, Fe-Np, Fe-W, Fe-Zn, Ge-Ni, La-Sn, La-Ti, La-Zr, Li-Sn, Mn-S, and Nb-Re. *J. Phase Equilibria Diffus.* **2016**, *37*, 621–634. [CrossRef]
55. Okamoto, H. Supplemental Literature Review of Binary Phase Diagrams: Au-La, Ce-Pt, Co-Pt, Cr-S, Cu-Sb, Fe-Ni, Lu-Pd, Ni-S, Pd-Ti, Si-Te, Ta-V, and V-Zn. *J. Phase Equilibria Diffus.* **2019**, *40*, 743–756. [CrossRef]
56. Wen, Z.; Wang, Y.; Wang, C.; Jiang, M.; Li, H.; Ren, Y.; Qin, G. Redetermination of the Fe–Pt phase diagram by using diffusion couple technique combined with key alloys. *Int. J. Mater. Res.* **2022**, *113*, 428–439. [CrossRef]
57. Wang, B.; Berry, D.C.; Chiari, Y.; Barmak, K. Experimental measurements of the heats of formation of Fe3Pt, FePt, and FePt3 using differential scanning calorimetry. *J. Appl. Phys.* **2011**, *110*, 013903. [CrossRef]

nanomaterials MDPI

Article

Significant Modulation of Vortex Resonance Spectra in a Square-Shape Ferromagnetic Dot

Shaojie Hu [1,2], Xiaomin Cui [3,*], Kang Wang [1], Satoshi Yakata [4] and Takashi Kimura [2,*]

1 Center for Spintronics and Quantum Systems, State Key Laboratory for Mechanical Behavior of Materials, School of Materials Science and Engineering, Xi'an Jiaotong University, Xi'an 710049, China; shaojiehu@mail.xjtu.edu.cn (S.H.); 3120102052@stu.xjtu.edu.cn (K.W.)
2 Department of Physics, Kyushu University, 744 Motooka, Fukuoka 819-0395, Japan
3 School of Physical Science and Technology, Northwestern Polytechnical University, Xi'an 710129, China
4 Department of Information Electronics, Fukuoka Institute of Technology, 3-30-1 Wajiro-higashi, Higashi-ku, Fukuoka 811-0295, Japan; yakata@fit.ac.jp
* Correspondence: sadycui@nwpu.edu.cn (X.C.); t-kimu@phys.kyushu-u.ac.jp (T.K.)

Abstract: The resonance property of a magnetic vortex contained within a micron-sized square Py dot was detected using an amplitude-modulated magnetic field excitation technique. We found a significant modulation of the resonant spectra as the external magnetic field changes. The Lorentzian-like spectrum changes from a peak to a dip via a transition of anti-Lorentzian-like spectra. By conducting the micromagnetic simulations, we confirmed that the transition behavior results from the unusual resistance change depending on the vortex core center position. Additionally, the power dependence of the anti-Lorentzian-like spectra revealed a fairly persistent coexistence of peak and dip. Thus, the tunable spectra suggest one way to develop an integratable radiofrequency microcircuits.

Keywords: magnetic vortex; vortex core resonance; spin dynamics; patterned ferromagnetic structure

Citation: Hu, S.; Cui, X.; Wang, K.; Yakata, S.; Kimura, T. Significant Modulation of Vortex Resonance Spectra in a Square-Shape Ferromagnetic Dot. *Nanomaterials* **2022**, *12*, 2295. https://doi.org/10.3390/nano12132295

Academic Editor: Julian Maria Gonzalez Estevez

Received: 15 June 2022
Accepted: 2 July 2022
Published: 4 July 2022

1. Introduction

The dynamic properties of magnetic materials or devices are interesting research areas because of their vast frequency tuning range from several megahertz to gigahertz and even terahertz [1–4]. The total energy of a ferromagnetic unit is determined by the competition among the exchange field, demagnetization field, external field and anisotropic field etc. [5]. Studies of ferromagnetic resonance or spin waves often focus on ferromagnetic materials with an external field that exceeds the anisotropic field and demagnetization field [2]. However, at low magnetic fields, the magnetic momentum distribution becomes nonuniform due to the competition between the anisotropic field and demagnetization field, which depend on the symmetry of the crystal structure and geometry, respectively. The complex magnetization distribution complicates magnetic dynamics analysis. The magnetic vortex, a topological structure defined by in-plane curling magnetization and out of core magnetization, has piqued the interest of researchers owing to its exceptional thermal stability [6–8]. The vortex core could be triggered to gyroscopic motion around its equilibrium point with a sub-gigahertz resonant frequency [9–12]. The study of vortex core dynamics would contribute to a better understanding of vortex resonant modes and the underlying physics, hence enabling the creation of vortex-based oscillators or filters. Much research has been conducted to investigate magnetic vortex dynamics in circular ferromagnetic disks [1,13–23], thanks to the simplified analysis in their perfect symmetric structures.

As is well-known, one effective way to manipulate the magnetic vortex property is to alter the geometry of the confined structure. A prior study demonstrated that asymmetric nucleation energy can be used to alter the chirality of an odd-sided polygon [24]. Additionally, the triangular dot with a magnetic vortex has been confirmed to have wide tunability with the external field [25,26]. Moreover, well-defined magnetic vortex excitation in square elements has been obtained by measuring the induction voltage while applying an external field along the edge [11,27]. To broaden our understanding, we applied a sensitive electrical measurement approach with the separation between the excitation and detection circuits that enables the neglect of complex analysis and the access to dynamics under high power excitation [28]. Therefore, in this study, we investigated the dynamics of a magnetic vortex confined in a chain of square ferromagnetic dots using an amplitude-modulated magnetic field excitation technique under a variety of external magnetic fields. The dynamic response spectrum transitioned from a Lorentzian-like shape to an anti-Lorentzian-like shape as the external field was changed. The transition process was thoroughly investigated and verified by conducting a micromagnetic simulation. Moreover, the power dependence of the anti-Lorentziant-like spectra has also been investigated.

2. Result and Discussion

A chain of square-shaped ferromagnetic dots was fabricated on Si substrate using conventional lift-off and electron beam lithography techniques. Here, a 40-nm-thick permalloy (Py) film was evaporated by electron-beam evaporation under the pressure of 2×10^{-7} Pa. The square device with a diagonal distance of 4 μm was fabricated. The diagonal distance corresponds to the diameter of the circumscribed circle of the square dot. Besides, the center-center distance was designed as 1.1 times of the diagonal distance. The Py dots were connected by Cu pads, and the microwave signal was injected into the Py dots by periodical Cu electrodes on top of the Py dots. Cu pads and electrodes with a thickness of 200 nm were deposited by a Joule heat evaporator after surface cleaning of the Py dots under low-energy Ar ion milling. In addition, SiO_2 was prepared to cut the connection between Py dots and Cu electrodes. As shown in the SEM image of part of the device in Figure 1b, one of the diagonal lines of the square is parallel to the chain of the Py dots.

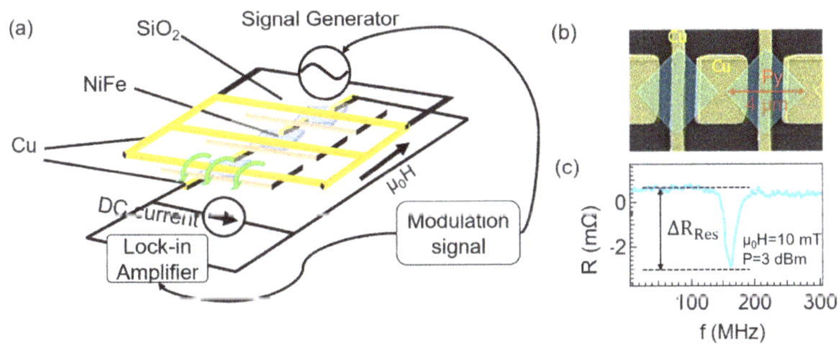

Figure 1. (**a**) Schematic measurement setup for the detection of magnetic vortex dynamics. An in-plane static magnetic field ($\mu_0 H$) is applied parallel with the chain of the Py dots. The amplitude-modulated RF field is injected from the periodically patterned Cu electrodes into the Py dots. The voltage is detected by flowing DC current in another separated circuit using a lock-in measurement system. (**b**) SEM image for the device with the diagonal distance of 4 μm and center-center distance of 4.4 μm of the square disk. (**c**) A representative spectrum as a function of the input RF frequency under a magnetic field of 10.0 mT. ΔR_{Res} is the resistance change between the oscillation and non-oscillation states.

To explore the dynamic properties of the magnetic vortices confined in square Py dots, a sensitive detection technique was employed and the schematic circuit is shown in Figure 1a. In the measurement, an amplitude-modulated RF signal was applied to the Cu electrode to oscillate the vortex core confined in the square dot. Simultaneously, the voltage response of the magnetic vortices was detected by another separated circuit combining the lock-in measurement system with a DC current flowing in the chain of Py dots. This separation between the excitation and detection line enables the simplification of the complex analysis. The dynamical properties of the magnetic vortices were observed by sweeping the RF frequency, while a static magnetic field was applied along the chain. As seen in Figure 1c, a typical spectrum with a resonant dip at 161.0 MHz was obtained at the static field of 10.0 mT. ΔR_{Res}, the difference between the baseline and the resonant dip, represents the magnetoresistance change between the oscillation and non-oscillation states of the magnetic vortices.

The resonant behavior was carefully investigated by changing the magnetic field. Figure 2a shows the image plot of frequency dependent spectra with sweeping the external magnetic field at RF power 3 dBm. First, we could see the resonant frequency varies from 98.0 MHz to 238.0 MHz, which indicates the achievement of a large modulation of the resonant frequency. At the low field region, the almost level white line indicates spectra have resonant peaks. However, the monotonically increased black line indicates the spectra have resonant dips in such field region. To clarity how does the transition of spectra, we also plot some spectra at specific fields in Figure 2b. Interestingly, we observed an significant anti-Lorentzian-like spectrum at 6.0 mT, where a resonant dip appears at 118.0 MHz besides the resonant peak of 98.0 MHz. The amplitude of the resonant peak is almost same as the resonant dip in this field. The resonant dip became stronger and dominant with the increase of field. Finally, it becomes a single resonant dip again when the magnetic field is larger than 7.0 mT. The experimental observation of the coexistence of both the oscillation peak and dip is unique. Usually, two oscillation responses (peak or dip) correspond to two oscillation states according to a previous study [29]. To clarify, we conducted a micromagnetic simulation to check whether these two states represent two oscillation modes or not.

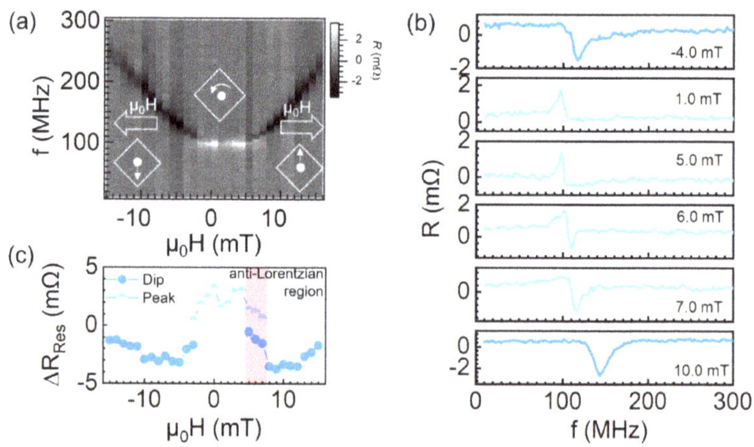

Figure 2. (**a**) The image plot of frequency dependent spectra with sweeping the external magnetic field. The vortex core displacement is also shown with counterclockwise chirality. (**b**) Magnetoresistance spectra with specific external magnetic fields. (**c**) ΔR_{Res} as a function of the magnetic fields. The blue and light blue dots stand for the resonant dip and peak, respectively.

Here, we performed the micromagnetic simulation using MuMax^3 [30]. The computational structure was designed with the same size and thickness as the fabricated Py dot. The domain was discretized into mesh sizes of 4 nm $\times$ 4 nm $\times$ 40 nm. We used the typical microstructural parameters of Permalloy, with an exchange stiffness constant $A_{ex} = 1.3 \times 10^{-11}$ J/m, damping parameter $\alpha = 0.006$, saturation magnetization $M_s = 8 \times 10^5$ A/m, and zero magnetocristalline anisotropy constant. During the simulation, we assumed that the chirality is counterclockwise and the polarity points vertically up of the plane. First, we obtained the in-plane components of the magnetization distribution

M confined in the square dot under various external fields parallel to the diagonal direction, as shown in Figure 3a. It is natural to see the vortex core shifts up under a positive field in the x-direction.

To obtain the resonant spectra, a sinc based exciting field, $h(t) = h_0 \mathrm{sinc}(2\pi f_c(t - t_0))\hat{e}_z$, with $\mu_0 h_0 = 10$ mT, $f_c = 50$ GHz, $t_0 = 5$ ns, is applied locally in the wave guide. Based on the present magnetization profile, we calculated the field dependence of the FFT spectra of magnetic momentum for vortex-core gyrations as shown in Figure 3b. The resonant frequencies are within the same range as the experimental results. For further comparison, we also give the image plot of the frequency spectra for the external field in Figure 3c. The continued white line indicates that each spectrum only has one resonant peak in all field regions. This implies that there should be just one oscillation mode, which would not explain the experimental spectrum transition.

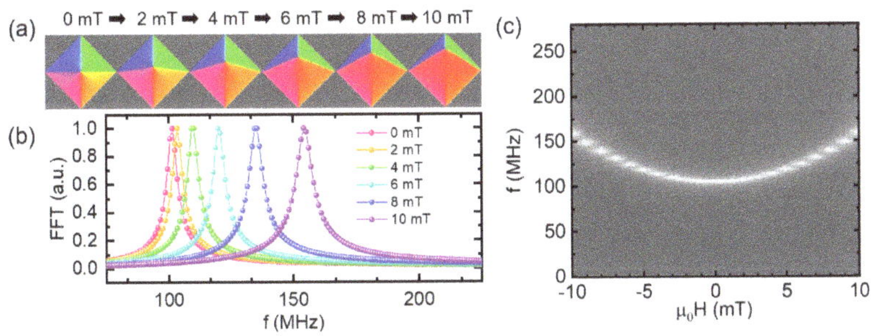

Figure 3. (**a**) Simulated domain structures of the magnetic vortex under various external magnetic fields. (**b**) Field dependence of the FFT spectra under the partial positive magnetic fields. (**c**) Image plot of the FFT spectra under the external magnetic from −10.0 mT to 10.0 mT.

To gain insight into the observed unique spectra, the dynamic processes were further studied by calculating the corresponding magnetoresistance in the Py dot. With assuming the current along the x direction, we can use the magnetization direction of each unit cell to calculate its resistivity using the anisotropic magnetoresistance (AMR) effect. The AMR of Py is given by the following equation:

$$\rho = \rho_0(1 + \chi cos^2\theta) \quad (1)$$

where, ρ_0, the resistivity of Py, is approximately $3.6\times10^8\ \Omega\cdot$m. Based on our experimental results, the AMR ratio χ is about 1.4%. θ is the angle between the magnetization and the current direction. Finally, the total resistance was calculated by connecting all cell resistance in parallel and series for each magnetic states.

Figure 4a provides a contour plot of the magnetoresistance as a function of the vortex core position in the square dot. It's clearly to see the homogeneously distributed magnetoresistance with two-fold symmetry. The magnetoresistance is largest at the maximum y value, whereas it is smallest at the maximum x value. Besides, the vortex core trajectory is circular shape at zero field during oscillation state. However, the core trajectory deviates from circular and the radius of the core motion decreases under 10.0 mT. It appears that the shape distorts significantly when the vortex core moves to the top. This should result from a larger restriction of the potential when the vortex core approaches the upper vertex. Correspondingly, the resonant frequency increases from zero field to 10.0 mT.

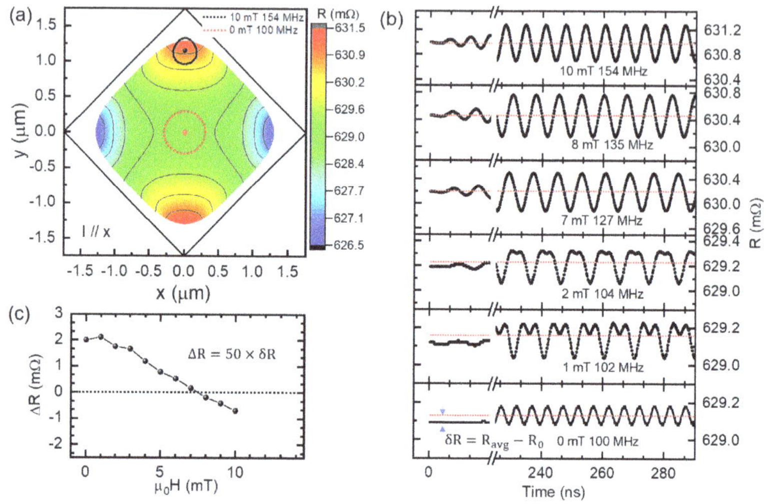

Figure 4. (**a**) Contour map of magnetoresistance in the square Py dot as a function of the vortex core center position. The magnetoresistance was calculated for current parallel to x direction. Besides, the core trajectories of resonant state are plotted as the red dotted line and black dotted line under 0 mT and 10.0 mT, respectively. (**b**) The numerically calculated time dependence of anisotropic magnetoresistance under the resonant state with the various external fields. (**c**) The calculated magnetoresistance change ΔR for 50 square dots with respect to the magnetic field.

In addition, we plotted the time dependent resistance at resonant frequency under some of the magnetic fields in Figure 4b. We also calculated the average resistance after the stable oscillation in ten periods and marked it as red dashed lines. By calculating the difference between the resistance in the non-oscillation and oscillation states, we obtained δR which can response the oscillation states. For comparing with the experimental results, we also plotted ΔR of 50 square dots as a function of the external field as shown in Figure 4c. The calculated value agrees with the experimental result within the same order of magnitude. Moreover, ΔR first increases with the external field and then decreases to the negative value. This tendency is also in good agreement with the experimental results shown in Figure 2c. To understand this behavior, we checked the time dependence of the resistance curve in Figure 4b. We can notice that the two-fold symmetry resistance response is broken when the field is away zero. Finally, only one-fold symmetry resistance response appears. During the transition, the significant deviation from the sine function of the curve indicates an unusual magnetoresistance change under this field during oscillation. So, the highest δR value was observed around 1.0 mT based on the numerical results in Figure 4c. This feature was also clearly observed in experimental results shown in Figure 2c.

As for the experimentally observed anti-Lorentzian-like spectra, we think that it should match a special magnetic field with both $\delta R > 0$ and $\delta R < 0$. According to Figure 4c, the ΔR value for 7.0 mT is positive, whereas that for 8.0 mT is negative. Correspondingly, we predicted this special situation located in the field range between 7.0 mT and 8.0 mT. Therefore, we further studied the time-dependent resistance for the field between 7.0 mT and 8.0 mT and calculated the δR with respect to the RF frequency. As shown in Figure 5a, we successfully obtained the special spectra with both the peak and dip at the magnetic field of 7.3 mT. Besides, we plotted the core trajectories of different frequencies under 7.3 mT. As seen in Figure 5b, the trajectory radius of 127.0 MHz is the largest compared with that of 125.0 MHz and 131.0 MHz. Those results indicate that the resonant frequency should be 127.0 MHz at 7.3 mT, even the corresponding ΔR is almost zero. Therefore, the experimentally obtained resistance response of the peak and dip in the spectrum should be attributed to the unusual magnetoresistance change of the core position during oscillation, rather than the two oscillation modes. Namely, the anti-Lorentzian resistance curve with the peak and dip is an extrinsic behavior of one vortex oscillation mode. By examining the simulated spectra obtained under 9 dBm and 12 dBm, we find the peak and dip are still very stable. In addition, ΔR increases with an increase in RF power.

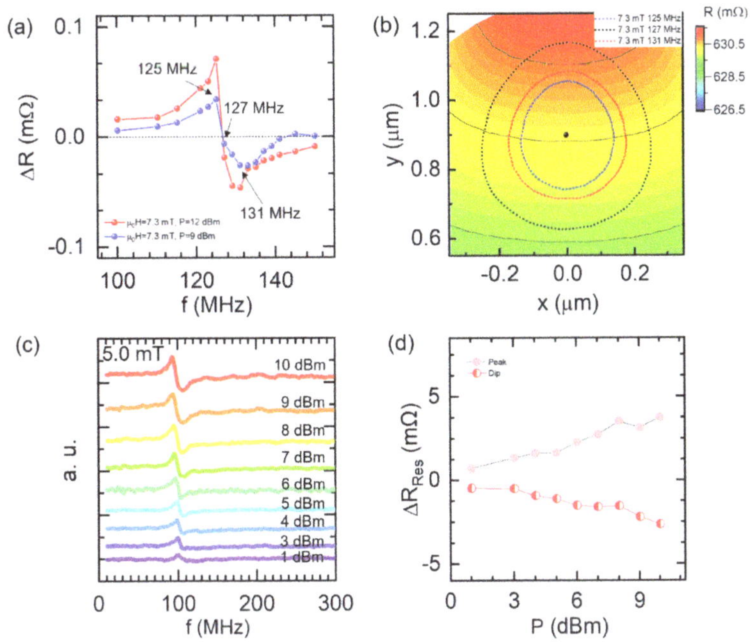

Figure 5. (**a**) The numerical resistance change as a function of the excitation frequency with the magnetic field of 7.3 mT at 9 and 12 dBm, respectively. (**b**) Simulated core trajectories at excitation frequency of 125 MHz, 127 MHz, and 131 MHz with the same magnetic field of 7.3 mT and RF power of 9 dBm. (**c**) The experimental spectra obtained with the magnetic field of 5.0 mT at various RF power. (**d**) ΔR_{Res} as a function of the RF power; The red dot and red half-filled dot represent the value of the resonant peak and resonant dip, respectively.

To validate this simulation result, we evaluated the dynamical behavior with a magnetic field of 5.0 mT under different RF power as shown in Figure 5c. Clearly, the double resonance responses of the device exist consistently from 1 dBm to 10 dBm, demonstrating that the anti-Lorentz-like spectra are very stable with increasing RF power. We also summarized the ΔR_{Res} for both the resonant peak and dip with respect to the RF power, as shown in Figure 5d. Both ΔR_{Res} monotonically increase, which is in good agreement with the simulation. Moreover, it seems that the upside resonant peak becomes weaker while the resonant dip becomes stronger and dominant with the power increase at 5.0 mT. It may be related to the slight resonant frequency modulation under higher excitation power.

3. Conclusions

We experimentally explored the resonant properties of a chain of square-shaped Py dots. A large modulation has been achieved not only for the resonant frequency but also for the shape of the resonance spectra. The spectra changes from a peak to a dip via an anti-Lorentzian transition with the external magnetic field. We confirmed that the oscillation peak and dip should be attributed to the unusual resistance change depending on the core-center position rather than two oscillation modes using the micromagnetic simulation. They may find an use in signal processing based on microwave logic circuits.

Author Contributions: Conceptualization, S.Y. and T.K.; Data curation, S.H. and X.C.; Formal analysis, X.C.; Methodology, S.H., X.C., S.Y. and T.K.; Software, K.W. and S.Y.; Supervision, T.K.; Writing—original draft, S.H. and X.C.; Writing—review & editing, S.H. and X.C. All authors have read and agreed to the published version of the manuscript..

Funding: This study is partially supported by National Natural Science Foundation of China for Young Scholar (51701158), National Key Research Program of China (Grant No. 2017YFA0206200), JSPS KAKENHI Grant Numbers 17H06227, 21H05021 and JST CREST (JPMJCR18J1).

Data Availability Statement: The data supporting the findings of this study are available from the corresponding author upon reasonable request.

Conflicts of Interest: The authors declare no conflict of interest.

References

1. Kasai, S.; Nakatani, Y.; Kobayashi, K.; Kohno, H.; Ono, T. Current-driven resonant excitation of magnetic vortices. *Phys. Rev. Lett.* **2006**, *97*, 107204. [CrossRef] [PubMed]
2. Demidov, V.E.; Urazhdin, S.; Liu, R.; Divinskiy, B.; Telegin, A.; Demokritov, S.O. Excitation of coherent propagating spin waves by pure spin currents. *Nat. Commun.* **2016**, *7*, 10446. [CrossRef] [PubMed]
3. Walowski, J.; Münzenberg, M. Perspective: Ultrafast magnetism and THz spintronics. *J. Appl. Phys.* **2016**, *120*, 140901. [CrossRef]
4. Bondarenko, P.; Sherman, E.Y. Uniform magnetization dynamics of a submicron ferromagnetic disk driven by the spin–orbit coupled spin torque. *J. Phys. D Appl. Phys.* **2017**, *50*, 265004. [CrossRef]
5. Antos, R.; Otani, Y.; Shibata, J. Magnetic vortex dynamics. *J. Phys. Soc. Jpn.* **2008**, *77*, 031004. [CrossRef]
6. Cowburn, R.P.; Koltsov, D.K.; Adeyeye, A.O.; Welland, M.E.; Tricker, D.M. Single-Domain Circular Nanomagnets. *Phys. Rev. Lett.* **1999**, *83*, 1042–1045. [CrossRef]
7. Shinjo, T. Magnetic Vortex Core Observation in Circular Dots of Permalloy. *Science* **2000**, *289*, 930–932. [CrossRef]
8. Lukyanchuk, I.; Vinokur, V.; Rydh, A.; Xie, R.; Milošević, M.; Welp, U.; Zach, M.; Xiao, Z.; Crabtree, G.; Bending, S.; et al. Rayleigh instability of confined vortex droplets in critical superconductors. *Nat. Phys.* **2015**, *11*, 21–25. [CrossRef]
9. Thiele, A.A. Steady-state motion of magnetic domains. *Phys. Rev. Lett.* **1973**, *30*, 230–233. [CrossRef]
10. Guslienko, K.Y.; Ivanov, B.A.; Novosad, V.; Otani, Y.; Shima, H.; Fukamichi, K. Eigenfrequencies of vortex state excitations in magnetic submicron-size disks. *J. Appl. Phys.* **2002**, *91*, 8037–8039. [CrossRef]
11. Park, J.P.; Eames, P.; Engebretson, D.M.; Berezovsky, J.; Crowell, P.A. Imaging of spin dynamics in closure domain and vortex structures. *Phys. Rev. B* **2003**, *67*, 020403. [CrossRef]
12. Guslienko, K.Y.; Lee, K.S.; Kim, S.K. Dynamic origin of vortex core switching in soft magnetic nanodots. *Phys. Rev. Lett.* **2008**, *100*, 027203. [CrossRef]
13. Buchanan, K.S.; Roy, P.E.; Grimsditch, M.; Fradin, F.Y.; Guslienko, K.Y.; Bader, S.D.; Novosad, V. Soliton-pair dynamics in patterned ferromagnetic ellipses. *Nat. Phys.* **2005**, *1*, 172–176. [CrossRef]
14. Novosad, V.; Fradin, F.Y.; Roy, P.E.; Buchanan, K.S.; Guslienko, K.Y.; Bader, S.D. Magnetic vortex resonance in patterned ferromagnetic dots. *Phys. Rev. B—Condens. Matter Mater. Phys.* **2005**, *72*, 024455. [CrossRef]
15. Shibata, J.; Nakatani, Y.; Tatara, G.; Kohno, H.; Otani, Y. Current-induced magnetic vortex motion by spin-transfer torque. *Phys. Rev. B* **2006**, *73*, 020403. [CrossRef]
16. Pribiag, V.S.; Krivorotov, I.N.; Fuchs, G.D.; Braganca, P.M.; Ozatay, O.; Sankey, J.C.; Ralph, D.C.; Buhrman, R.A. Magnetic vortex oscillator driven by dc spin-polarized current. *Nat. Phys.* **2007**, *3*, 498–503. [CrossRef]
17. Yamada, K.; Kasai, S.; Nakatani, Y.; Kobayashi, K.; Kohno, H.; Thiaville, A.; Ono, T. Electrical switching of the vortex core in a magnetic disk. *Nat. Mater.* **2007**, *6*, 270–273. [CrossRef]
18. Gaididei, Y.; Sheka, D.D.; Mertens, F.G. Controllable switching of vortex chirality in magnetic nanodisks by a field pulse. *Appl. Phys. Lett.* **2008**, *92*, 012503. [CrossRef]
19. Vogel, A.; Drews, A.; Kamionka, T.; Bolte, M.; Meier, G. Influence of Dipolar Interaction on Vortex Dynamics in Arrays of Ferromagnetic Disks. *Phys. Rev. Lett.* **2010**, *105*, 037201. [CrossRef]
20. Vogel, A.; Kamionka, T.; Martens, M.; Drews, A.; Chou, K.W.; Tyliszczak, T.; Stoll, H.; Van Waeyenberge, B.; Meier, G. Coupled vortex oscillations in spatially separated permalloy squares. *Phys. Rev. Lett.* **2011**, *106*, 137201. [CrossRef]
21. Sugimoto, S.; Fukuma, Y.; Kasai, S.; Kimura, T.; Barman, A.; Otani, Y. Dynamics of coupled vortices in a pair of ferromagnetic disks. *Phys. Rev. Lett.* **2011**, *106*, 197203. [CrossRef] [PubMed]
22. Dussaux, A.; Khvalkovskiy, A.V.; Bortolotti, P.; Grollier, J.; Cros, V.; Fert, A. Field dependence of spin-transfer-induced vortex dynamics in the nonlinear regime. *Phys. Rev. B—Condens. Matter Mater. Phys.* **2012**, *86*, 014402. [CrossRef]
23. Suess, D.; Bachleitner-Hofmann, A.; Satz, A.; Weitensfelder, H.; Vogler, C.; Bruckner, F.; Abert, C.; Prügl, K.; Zimmer, J.; Huber, C.; et al. Topologically protected vortex structures for low-noise magnetic sensors with high linear range. *Nat. Electron.* **2018**, *1*, 362–370. [CrossRef]
24. Yakata, S.; Miyata, M.; Nonoguchi, S.; Wada, H.; Kimura, T. Control of vortex chirality in regular polygonal nanomagnets using in-plane magnetic field. *Appl. Phys. Lett.* **2010**, *97*, 222503. [CrossRef]

25. Vogel, A.; Corinna Niemann, A.; Stenner, C.; Drews, A.; Im, M.Y.; Fischer, P.; Meier, G. Vortex dynamics in triangular-shaped confining potentials. *J. Appl. Phys.* **2012**, *112*, 063916. [CrossRef]
26. Yakata, S.; Tanaka, T.; Kiseki, K.; Matsuyama, K.; Kimura, T. Wide range tuning of resonant frequency for a vortex core in a regular triangle magnet. *Sci. Rep.* **2013**, *3*, 3567. [CrossRef]
27. Langner, H.H.; Bocklage, L.; Matsuyama, T.; Meier, G. Inductive detection of magnetic vortex gyration. *Phys. Rev. B-Condens. Matter Mater. Phys.* **2013**, *87*, 064420. [CrossRef]
28. Cui, X.; Hu, S.; Hidegara, M.; Yakata, S.; Kimura, T. Sensitive detection of vortex-core resonance using amplitude-modulated magnetic field. *Sci. Rep.* **2015**, *5*, 17922. [CrossRef]
29. Kuepferling, M.; Serpico, C.; Pufall, M.; Rippard, W.; Wallis, T.M.; Imtiaz, A.; Krivosik, P.; Pasquale, M.; Kabos, P. Two modes behavior of vortex oscillations in spin-transfer nanocontacts subject to in-plane magnetic fields. *Appl. Phys. Lett.* **2010**, *96*, 252507. [CrossRef]
30. Vansteenkiste, A.; Leliaert, J.; Dvornik, M.; Helsen, M.; Garcia-Sanchez, F.; Van Waeyenberge, B. The design and verification of MuMax3. *AIP Adv.* **2014**, *4*, 107133. [CrossRef]

nanomaterials

MDPI

Article

Characterization of Mn_5Ge_3 Contacts on a Shallow Ge/SiGe Heterostructure

Troy A. Hutchins-Delgado [1,2,3,*], Sadhvikas J. Addamane [1,2], Ping Lu [1] and Tzu-Ming Lu [1,2]

1 Sandia National Laboratories, Albuquerque, NM 87185, USA; saddama@sandia.gov (S.J.A.); plu@sandia.gov (P.L.); tlu@sandia.gov (T.-M.L.)
2 Center for Integrated Nanotechnologies, Albuquerque, NM 87185, USA
3 Center for High Technology Materials, University of New Mexico, Albuquerque, NM 87106, USA
* Correspondence: tahutch@sandia.gov

Abstract: Mn_5Ge_3 is a ferromagnetic phase of the Mn-Ge system that is a potential contact material for efficient spin injection and detection. Here, we investigate the creation of Mn_5Ge_3-based contacts on a Ge/SiGe quantum well heterostructure via solid-state synthesis. X-ray diffraction spectra fitting indicates the formation of Mn_5Ge_3-based contacts on bulk Ge and Ge/SiGe. High-resolution scanning transmission electron microscopy imaging and energy dispersive X-ray spectroscopy verify the correct Mn_5Ge_3-based phase formation. Schottky diode measurements, transmission line measurements, and Hall measurements reveal that Mn_5Ge_3-based contacts serve as good p-type contacts for Ge/SiGe quantum well heterostructures due to having a low Schottky barrier height of 0.10 eV (extracted from a Mn_5Ge_3/n-Ge analogue) and a contact resistance in the order of 1 kΩ. Furthermore, we show that these electrical characteristics have a gate-voltage dependence, thereby providing tunability.

Keywords: Mn_5Ge_3; Ge/SiGe; thin film; germanide; phase formation; solid-state synthesis

Citation: Hutchins-Delgado, T.A.; Addamane, S.J.; Lu, P.; Lu, T.-M. Characterization of Mn_5Ge_3 Contacts on a Shallow Ge/SiGe Heterostructure. *Nanomaterials* **2024**, *14*, 539. https://doi.org/10.3390/nano14060539

Academic Editor: Dong-Joo Kim

Received: 26 January 2024
Revised: 14 March 2024
Accepted: 15 March 2024
Published: 19 March 2024

1. Introduction

Holes in germanium (Ge) have been shown to be an excellent candidate for spin-based alternative computing technologies, such as spin field-effect transistors (SpinFETs) for classical computing [1–3] and hole spin qubits for quantum computing [4–10]. In either technology, it is paramount to start with the best possible material. Recently, there have been demonstrations of Ge/SiGe quantum wells with mobilities in the 1×10^6 $cm^2\,V^{-1}\,s^{-1}$ range [11,12]. Large-scale wafer growth of high-quality Ge/SiGe quantum materials has now opened access for further research [12,13].

Germanium has an advantage whereby Rashba spin-orbit coupling (SOC) is dominant over Dresselhaus SOC, thereby allowing for gate-tunable SOC. A gate-tunable SOC is crucial for SpinFET operation because it allows for direct control over the spin precession rate. However, control over spin precession is challenging if the transport is diffusive within the spin-orbit length. Fortunately, shallow and undoped germanium quantum wells have been shown to meet these requirements [14]. For SpinFETs, it is also important to have good spin injection and detection efficiencies. Optical spin injection in Ge/SiGe quantum wells has been theoretically shown to have 96% spin polarization [15] and experimentally shown to have 85% spin polarization [16]. While optical spin injection has high degrees of spin polarization, it is simply not trivial to extend to integrated systems for spintronic applications as compared with electrical spin injection.

Electrical spin injection is usually performed with a ferromagnetic contact that has either a conductivity mismatch with the substrate [17,18] or a tunnel junction [19–23]. Mn_5Ge_3 has been theoretically [24] and experimentally shown to be a good ferromagnetic contact for spin injection and detection in both p-type [1,2] and n-type [25] Ge. The easiest method to form Mn_5Ge_3 contacts is through solid-state synthesis whereby Mn thin films are deposited on Ge and annealed to form Mn_5Ge_3. However, to the best of our knowledge,

Mn_5Ge_3 solid-state synthesis has not been directly applied to strained Ge/SiGe quantum well heterostructures. Here, we study the formation of Mn_5Ge_3-based contacts on Ge/SiGe through structural characterization and evaluate their electrical performance. Our study reveals the complexity of a solid-state reaction that can occur with forming Mn_5Ge_3 directly on a Ge/SiGe heterostructure. Our unique results thus provide a foundation for future studies regarding the solid-state synthesis of Mn_5Ge_3 contacts for spin injection/detection on high-quality Ge/SiGe material.

2. Materials and Methods

Mn thin films were deposited onto 1 cm^2 die of bulk (100) Ge and Ge/SiGe heterostructures via thermal evaporation of Mn powder immediately after surface cleaning. The surface cleaning procedure consists of standard solvent clean, oxygen plasma clean, and three cycles of buffered oxide etchant (BOE) and deionized water (DI H_2O) rinse [26]. The Ge/SiGe heterostructures used consist of a 15 nm Ge quantum well at a depth of 30 nm with $Si_{0.15}Ge_{0.85}$ barriers and a Si capping layer with an approximate thickness of 1 nm. Details about the growth and transport properties of the Ge/SiGe heterostructure have been previously reported [13]. The deposition of approximately 100 nm of Mn occurred at a pressure of 1×10^{-6} torr and a rate of 0.5 Å s^{-1}. The die was then annealed at various temperatures for 30 min in a flash lamp annealer with Ar atmosphere. A Mn thickness of 100 nm was selected to ensure that there was enough Mn for the solid-state reaction to reach past the quantum well.

3. Results and Discussion

3.1. X-ray Diffraction Analysis

Mn thin films on bulk Ge were investigated first to verify that our process could result in Mn_5Ge_3 phase forming before proceeding to the Ge/SiGe heterostructures. Studies have shown that Mn_5Ge_3 formed via solid-state synthesis on (100) Ge and can occur from 250 °C up to 450 °C [27,28]. Therefore, we chose to proceed with 50 °C increments from 200 °C to 400 °C to deduce the appropriate annealing temperature for our annealing setup. X-ray diffraction (XRD) analysis of $\theta - 2\theta$ scans was performed to investigate the various phases of the Mn-Ge system formed at each of the annealing temperatures. The resulting $\theta - 2\theta$ scans and their analyses are shown in Figure 1. A detailed summary of the XRD analysis is represented in Table 1. Table 1 highlights the various phases found with corresponding 2θ, planes, lattice parameters (d-values), unit cell volume, and crystallite sizes extracted from the prominent peak of each phase at each annealing temperature. Crystallite sizes were determined using the Scherrer equation:

$$\tau = 0.94\frac{\lambda}{\beta cos(\theta)}. \tag{1}$$

Here, τ is the crystallite size in Å, λ is the X-ray wavelength in Å, β is the full width at half max in radians, and θ is the diffraction angle. The XRD system we use has a wavelength of 1.540 59 Å.

The XRD results for Mn on bulk Ge are summarized in Figure 1a, and fitting was performed with the International Centre for Diffraction Data (ICDD) database, PDF-2 [29]. When annealed at 200 °C (black curve), the film was determined to be a Mn-α phase. The Mn-α phase's characteristic feature is the peak at 43° corresponding to the (411) plane. Annealing at 200 °C resulted in a Mn-α crystallite size of 171.08 Å. Similar to annealing at 200 °C, annealing at 250 °C (red curve) and 300 °C (blue curve) also resulted in the Mn-α phase with crystallite sizes of 186.24 Å and 221.95 Å, respectively. However, at 300 °C, the peak around 43° started to diminish, and smaller signatures started to appear at both higher and lower diffraction angles. These observed deviations indicate that Mn started to react with Ge to form Mn_5Ge_3 as this is the first phase known to form [27,28]. Next, the sample annealed at 350 °C (green curve) showed the same signatures that started to appear in the 300 °C–anneal sample with higher intensities. Furthermore, the characteristic peak of the Mn-α phase at 43° disappeared and was replaced by two peaks at around 42.35° and 43.73° corresponding to the Mn_5Ge_3 (211) and (112) planes, respectively. Additionally, there were peaks at 35.46°, 38.24°, 38.38°, and 46.22° that were only prominent at 350 °C, which arose from the (002), (210), (102), and (202) Mn_5Ge_3 planes, respectively. The peaks of the 350 °C sample fit to Mn_5Ge_3, confirming that Mn_5Ge_3 indeed started to form at 300 °C. The prominent peak for Mn_5Ge_3 was at 42.35°, by which crystallite sizes of 376.85 Å, 370.46 Å, and 512.94 Å were extracted, respectively, at 300 °C, 350 °C, and 400 °C. By 400 °C (magenta curve), most Mn_5Ge_3 peaks disappeared, and the new peaks fit to $Mn_{11}Ge_8$. A crystallite size of 333.90 Å was obtained from the $Mn_{11}Ge_8$ prominent peak at 42.73°. Based on the XRD results above, these annealing conditions can be transferred to Ge/SiGe heterostructures to form Mn_5Ge_3 and likely within the narrower range of 300–400 °C with the most promising temperature being 350 °C.

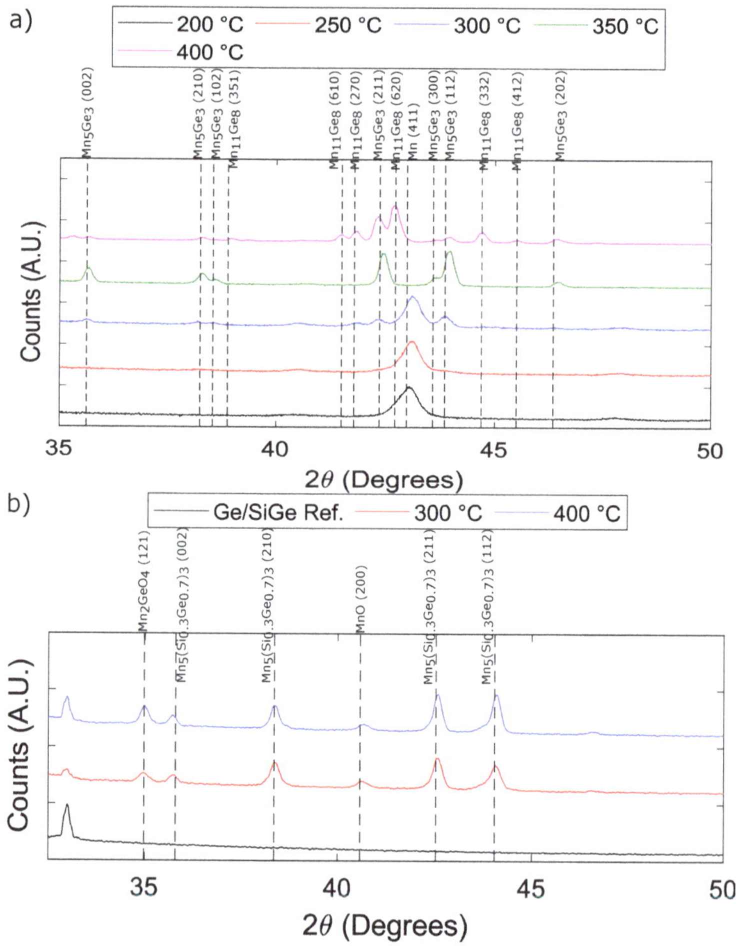

Figure 1. Summary of XRD $\theta - 2\theta$ scans. (**a**) XRD $\theta - 2\theta$ scans for thermal evaporated Mn thin films on (100) Ge substrates and annealed at 200 °C (black), 250 °C (red), 300 °C (blue), 350 °C (green), and 400 °C (magenta). The black vertical lines correspond to ICDD peak locations for Mn_5Ge_3 and $Mn_{11}Ge_8$. (**b**) XRD $\theta - 2\theta$ scans for thermal evaporated Mn thin films on Ge/$Si_{0.15}Ge_{0.85}$ quantum well heterostructures and annealed at 300 °C (red) and 400 °C (blue). The Ge/$Si_{0.15}Ge_{0.85}$ reference scan is shown in black. The vertical lines correspond to ICDD peak locations for Mn_5 $(Si_{0.3}Ge_{0.7})_3$ and various oxides: Mn_2GeO_4 and MnO.

Table 1. Table summary of X-ray diffraction analysis.

Phase	2θ	Plane (h k l)	d-Value (Å)	Unit Cell Volume ($Å^3$)	Crystallite Size (200 °C)	Crystallite Size (250 °C)	Crystallite Size (300 °C)	Crystallite Size (350 °C)	Crystallite Size (400 °C)
Mn-α	43.02°	4 1 1	2.201	707.85	171.08 Å	186.24 Å	221.95 Å	-	-
MnO	40.55°	2 0 0	2.223	87.88	-	-	271.51 Å	-	264.48 Å
Mn_2GeO_4	34.74°	1 2 1	2.580	337.59	-	-	314.93 Å	-	359.30 Å
Mn_5Ge_3	35.46°	0 0 2	2.530	226.114	-	-	-	-	-
Mn_5Ge_3	38.24°	2 1 0	2.352	226.114	-	-	-	-	-
Mn_5Ge_3	38.38°	1 0 2	2.343	226.114	-	-	-	-	-
Mn_5Ge_3	42.35°	2 1 1	2.132	226.114	-	-	376.85 Å	370.46 Å	512.94 Å
Mn_5Ge_3	43.61°	3 0 0	2.074	226.114	-	-	-	-	-
Mn_5Ge_3	43.73°	1 1 2	2.066	226.114	-	-	-	-	-
Mn_5Ge_3	46.22°	2 0 2	1.963	226.114	-	-	-	-	-
$Mn_5(Si_{0.3}Ge_{0.7})_3$	35.80°	0 0 2	2.487	218.45	-	-	-	-	-
$Mn_5(Si_{0.3}Ge_{0.7})_3$	38.37°	2 1 0	2.331	218.45	-	-	-	-	-
$Mn_5(Si_{0.3}Ge_{0.7})_3$	42.53°	2 1 1	2.111	218.45	-	-	317.68 Å	-	357.84 Å
$Mn_5(Si_{0.3}Ge_{0.7})_3$	44.03°	1 1 2	2.039	218.45	-	-	-	-	-
$Mn_{11}Ge_8$	39.32°	3 5 1	2.290	1055.77	-	-	-	-	-
$Mn_{11}Ge_8$	41.49°	6 1 0	2.175	1055.77	-	-	-	-	-
$Mn_{11}Ge_8$	42.26°	2 7 0	2.137	1055.77	-	-	-	-	-
$Mn_{11}Ge_8$	42.73°	6 2 0	2.115	1055.77	-	-	-	-	333.90 Å
$Mn_{11}Ge_8$	44.69°	3 3 2	2.026	1055.77	-	-	-	-	-
$Mn_{11}Ge_8$	45.50°	4 1 2	1.992	1055.77	-	-	-	-	-

Figure 1b shows the XRD results for the annealed Mn films on the Ge/SiGe heterostructure and a reference scan (black curve) of Ge/SiGe. The Ge/SiGe heterostructures were annealed at 300 °C (red curve) and 400 °C (blue curve). Both the 300 °C and 400 °C data are similar with only the relative heights of peaks changing between the two. The XRD peaks best fit to Mn_5 $(Si_{0.3}Ge_{0.7})_3$, Mn_2GeO_4, and MnO. However, the Mn_5 $(Si_{0.3}Ge_{0.7})_3$ peaks do not perfectly align with the data. We expect phases to have the form Mn_y $(Si_{0.15}Ge_{0.85})_x$ as our material has $Si_{0.15}Ge_{0.85}$ barriers, and this could explain the deviation from the fit. Nevertheless, XRD has detected the formation of the correct crystal structure type Mn_5 $(Si_xGe_{1-x})_3$, but microscopic verification is needed to confirm specifics about relative concentrations.

3.2. Scanning Transmission Electron Microscopy Analysis

We chose 350 °C as the optimal annealing temperature for making Mn_y $(Si_{0.15}Ge_{0.85})_x$ contacts since it was confirmed to be the best temperature for Mn_5Ge_3 on bulk Ge and gave a ±50 °C tolerance to the Ge/SiGe XRD results. We proceeded to make transmission line measurement (TLM) and Hall bar devices with Mn_y $(Si_{0.15}Ge_{0.85})_x$ contacts. An optical microscopy image in Figure 2a shows an example of the TLM (left) and Hall bar (right) devices after fabrication. The sample was used in electrical characterizations and microanalysis. Here, we discus the microanalysis results. High-resolution scanning transmission electron microscopy (HR-STEM) imaging, in combination with energy dispersive X-ray spectroscopy (EDS), was used to investigate the Mn_y $(Si_{0.15}Ge_{0.85})_x$ contacts in further detail. EDS provides a means to map elements in a structure by their respective Kα, Lα, and other X-ray signatures. The elements' signatures are then represented by a false coloring in the STEM images. The STEM high-angle annular dark-field (HAADF) image in Figure 2b reveals a complex structure stack with not-well-defined boundaries. The regions labeled I, II, and III correspond to the contact structure that is of interest. The material below the region of interest is the $Si_{0.15}Ge_{0.85}$ virtual substrate. The material atop the regions of interest correspond to the Au/Ti metal contact pads and the Al_2O_3 gate oxide. The EDS data in Figure 2c verify the clear distinction between the metal/oxide region and the complex contact regions. Here, Au is represented with green, blue corresponds to Ti, and red corresponds to Al (from Al_2O_3). Black represents the absence of any detected signatures. Figure 2d shows an overlay of Mn, O, and Ge maps, revealing the complexity of the structure. Here, green corresponds to Mn, blue is Ge, and red is O.

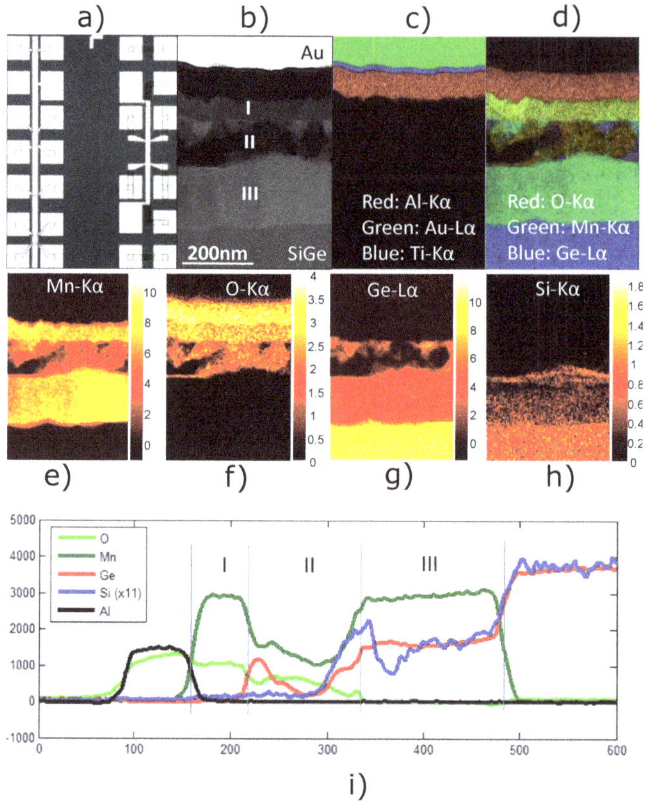

Figure 2. Cross-sectional STEM and EDS analysis of Mn_5Ge_3-based contacts for Ge/SiGe H-FET devices. (**a**) Optical microscope image TLM (**left**) and Hall bar (**right**) devices. (**b**) HAADF STEM image of the Mn_5Ge_3-based contact material stack. (**c**) Combined EDS map of the Au (red)/Ti (blue) gate metal and Al_2O_3 (Al-red) gate dielectric above the Mn_5Ge_3-based contact. (**d**) Combined EDS mapping of O (red), Mn (green), and Ge (blue). (**e**) EDS map of Mn, (**f**) EDS map of O, (**g**) EDS map of Ge, (**h**) EDS mapping of Si, and (**i**) integrated EDS line profile with signatures of O (light green), Mn (dark green), Ge (red), Si (blue), and Al (black).

Focusing on region I, the Mn (Figure 2e) and O (Figure 2f) elemental maps show that region I is entirely composed of Mn and O and is approximately 70 nm thick. This claim is supported by the absence of Ge and Si in region I, as shown in their respective elemental maps in Figure 2g,h. The Mn and O maps provide supporting evidence that the broad peak in the XRD data indeed corresponds to oxidized Mn. Furthermore, region I was determined to be polycrystalline MnO through HR-STEM HAADF imaging. Region II, like region I, contains Mn and O. However, region II also contains Ge and is approximately 116 nm thick. Unlike any of the other regions, region II does not have a uniform distribution of elements. Together, the Ge (Figure 2g) and Si (Figure 2h) data highlight the complexity of region II. The Si map shows only trace amounts where there are large concentrations of Ge. The Ge map shows that there are islands of Ge mostly near the region I/region II interface. Interestingly, the Mn and O distributions are nearly uniform in region II except at the areas of highest Ge concentrations and where the EDS data do not correspond to any of the elements tested (black region). This black region, most easily visualized in Figure 2d, could indicate the presence of a void since there are none of the expected elemental signatures and a lack of resolvable features in the HAADF STEM image. The Mn and Ge maps indicate a low concentration likely coming from behind the void. While voids could still occur due to the diffusion of Ge as discussed, the formation here is likely due to the complexity of the reaction that occurred. A better-quality film will form when annealed in a high-vacuum environment, and voids will be less likely to form [30–33]. Besides the islands of Ge, region II is likely a mixture of MnO and $Mn_5Ge_3O_x$ due to its location between region I and region III, which is consistent with previous reports of solid-state syntheses of Mn_5Ge_3 [33–36].

Region III, the largest of the three regions with a thickness of approximately 186 nm, appears to have the best film quality (as determined by HR-STEM) and a promising elemental map for Mn_y $(Si_{0.15}Ge_{0.85})_x$ contacts. The film has relatively uniform distributions of Mn, Ge, and dilute Si, as emphasized by the integrated EDS line profile in Figure 2i. Furthermore, the Si line is magnified such that when Si and Ge counts are equal, it represents the relative concentration of the barrier $Si_{0.15}Ge_{0.85}$. The region II/III interface is where the Ge/SiGe heterostructure starts. While it is clear that Mn is being driven into the heterostructure, the presence of Ge in region II indicates that Ge has left the heterostructure, thus making the heterostructure more Si rich.

To further investigate the film, HR-STEM HAADF imaging along with its fast Fourier transform (FFT) pattern was used to determine the exact phase of Mn_y $(Si_{0.15}Ge_{0.85})_x$. The STEM HAADF image of Mn_y $(Si_{0.15}Ge_{0.85})_x$ in Figure 3a, along with its FFT pattern in Figure 3b, is in excellent agreement with the projected structure (Figure 3c) and simulated selected area electron diffraction (SAED) pattern (Figure 3d) corresponding to the hexagonal crystal structure of Mn_5Ge_3. The SAED pattern is calculated using reciprocal lattice and associated electron structure factors under the kinematical approximation [37]. The geometry of the pattern (or relative positions of the reciprocal lattice) is important in this context, and fits exactly the FFT pattern in [1, −2, 0], confirming the phase. Since the starting material consisted of mostly $Si_{0.15}Ge_{0.85}$ layers, except for the Ge quantum well and Si cap, it is likely that the relative concentrations of Si and Ge remained near their original concentrations such that the film is Mn_5 $(Si_{0.15}Ge_{0.85})_3$. The EDS line scan, calibrated to the $Si_{0.15}Ge_{0.85}$ virtual substrate beyond region III, is consistent with this claim. The XRD data showed a fit with $Mn_5Si_{0.9}Ge_{2.1}$ or Mn_5 $(Si_{0.3}Ge_{0.7})_3$, and this could be due to the larger concentration of Si at the interface of region II and region III. Nevertheless, our data revealed the correct phase of Mn_5 $(Si_xGe_{1-x})_3$ to make a p-type ferromagnetic contact for Ge/SiGe heterostructure field-effect transistors (HFETs).

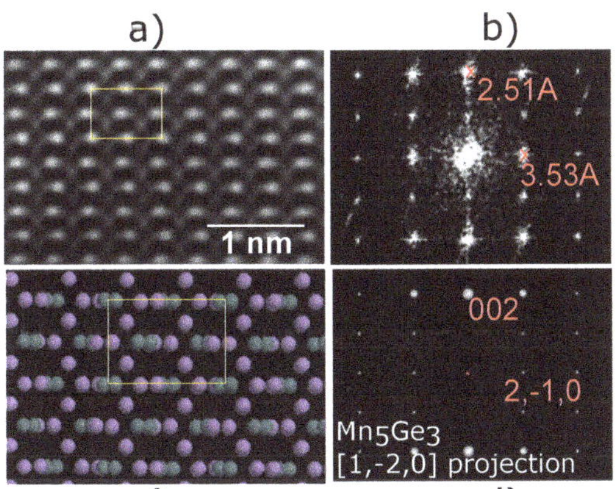

Figure 3. STEM HAADF analysis of Mn_5Ge_3-based contact. (**a**) HR-STEM HAADF of region III. (**b**) FFT of the STEM HAADF image. (**c**) Crystal structure of Mn_5Ge_3. (**d**) Simulated SAED pattern of Mn_5Ge_3 in the [1, −2, 0] projection.

3.3. Electrical Characterization

With the confirmation of solid-state-reaction-formed Mn_5Ge_3-based contacts on undoped Ge/SiGe heterostructures, we present the electrical properties of the Mn_5Ge_3 contacts from Schottky diodes, TLM, and Hall bar measurements. Since it can be difficult to analyze Schottky contacts on quantum well heterostructures, our Schottky diodes were fabricated on n-type (100) Ge, with a resistivity range of 1–10 Ωcm, to estimate the Schottky barrier height of Mn_5Ge_3/Ge. Mn_5Ge_3/n-Ge Schottky diodes serve as a good analogue for estimating the barrier height for Mn_5Ge_3/Ge in our Ge/SiGe heterostructure devices. Ideally, there would be an increase in barrier height due to the compressive strain of the Ge quantum well moving both the conduction and valence band edges farther apart [38,39]. However, Ge is widely known to pin the Fermi levels of most metals very close to the valence band edge, thus nullifying any effects due to strain in the case of hole transport. The lack of strain affecting barrier heights due to Fermi level pinning has been shown in the case of n-type SiGe [40]. While Mn_5Ge_3/n-Ge is not a direct replacement for the Mn_5Ge_3/stained-Ge quantum well, it does provide the best estimate available. A direct measurement for our system would require contact only of the quantum well with Mn_5Ge_3 in an isolated fashion, which is extremely difficult, especially given its relatively small thickness. To fabricate the Mn_5Ge_3/Ge Schottky diodes, Mn films on Ge were etched back into circular contacts and then annealed to form Mn_5Ge_3. The Schottky diodes were made with a radius of 200 µm. The Mn_5Ge_3/Ge Schottky diodes were characterized using current density–voltage analysis at room temperature. The current density–voltage data are represented in Figure 4a. The Schottky barrier height is extracted from the following relation:

$$\frac{k_B T}{e} \ln\left(\frac{J}{A^* T^2}\right) \approx \frac{V}{n} - \frac{\phi}{e}. \tag{2}$$

Here, $\frac{k_B T}{e}$ is the voltage corresponding to the thermal energy in eV, A^* is the Richardson constant, T is the temperature, J is the current density, V is the voltage, n is a nonideality factor, and $\frac{\phi}{e}$ is the Schottky barrier height. The red line in Figure 4d is the fit line where a Schottky barrier height, with respect to the Ge conduction band, of 0.57(1) eV is obtained. The fit assumes $A^* = 44.5(105)\ \mathrm{A\,cm^{-2}\,K^{-2}}$ and $T = 300\ \mathrm{K}$. Therefore, Mn_5Ge_3 has a Schottky barrier height of 0.10(1) eV with respect to the Ge valence band. The assumptions made and the obtained Schottky barrier height are consistent with other studies [1,2,25,32,41].

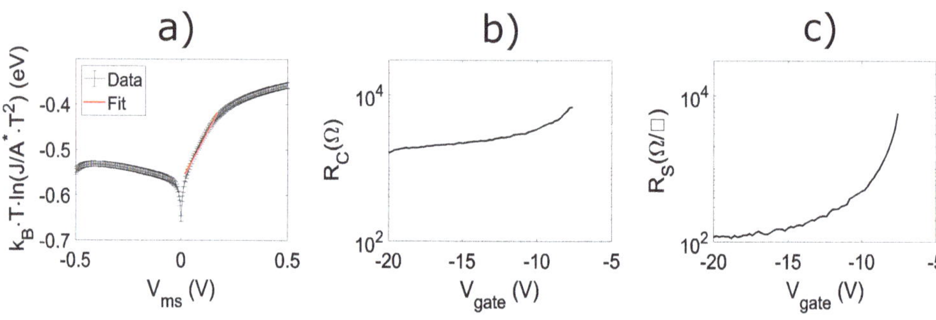

Figure 4. Electrical characterization of Mn_5Ge_3 contacts. (**a**) Mn_5Ge_3/n-type (100) Ge Schottky diode. (**b**) Contact resistance for TLM H-FET device. (**c**) Sheet resistance for TLM H-FET device.

While the Schottky diode on bulk n-type Ge gives an estimate of the Mn_5Ge_3/Ge barrier height for the heterostructure, it does not provide a full scope of the contact's electrical characteristics. Devices made with Ge/SiGe will be enhancement-mode heterostructure field-effect transistors (HFETs), whose characteristics will be dependent on applied gate voltages. Furthermore, the material was grown such that there was high-quality transport at cryogenic temperatures. Therefore, we fabricated TLM and Hall bar devices (as previously described [13], except with surface preparation, as described in this manuscript) to measure the Mn_5Ge_3 contacts' properties under these conditions. TLM and Hall measurements were performed using four-terminal, low-frequency lock-in techniques at a temperature of 4 K using a liquid helium cryostat. A 100 µV sinusoidal excitation was sourced from a lock-in for both the TLM and Hall measurements. For Hall measurements, the magnetic field was swept from −0.5 to 0.5 T.

TLM devices provide a means to extract the contact and gate-induced 2D hole gas (2DHG) sheet resistances of materials. Our TLM devices (left device in Figure 2a) were made with 25 µm × 50 µm Mn_5Ge_3–based contacts with separations (channel lengths) of 50 µm, 100 µm, 325 µm, 400 µm, and 675 µm. A 50 µm–wide gate that overlaps with all the contacts provides a means to capacitively induce 2DHG. A more negative gate voltage increases the 2DHG density and decreases the 2DHG sheet resistance. Contact resistance was obtained from the TLM device using two different methods, each using the TLM equation, assuming that the contact resistance is much larger than the resistance of the metal pads:

$$R_T = R_S \frac{L}{w} + 2R_C. \tag{3}$$

Here, R_T is the total measured resistance, R_S is the 2DHG sheet resistance, $\frac{L}{w}$ is the ratio of the channel length to the channel width, and R_C is the contact resistance. The first method used is the standard approach, where the total resistance is measured as a function channel length, a linear fit is performed, and the contact resistance is half the y-intercept of the linear fit. The 2DHG sheet resistance is then extracted by multiplying the fit's slope by the channel width. The extracted contact resistance and 2DHG sheet resistance, as a function of gate voltage, are shown in Figure 4b,c, respectively. A contact resistance in the order of 1 kΩ across the measured voltage range was obtained. Interestingly, the 2DHG sheet resistance saturates at gate voltages more negative than −15 V, while the contact resistance has a linear dependence on gate voltage after the saturation of the device. The gate voltage dependence past saturation can be explained by the gate's overlap with the Mn_5Ge_3-based contact. Although the saturation of holes has been achieved in the channel of the device, the electric field from the gate lowers the injection barrier. Thus, as the gate voltage is further increased past saturation, there is still a measured reduction in the contact resistance [42–44], while the 2DHG sheet resistance is no longer sensitive to the gate voltage. If there was no Schottky barrier at the Mn_5Ge_3/Ge interface in the Ge/SiGe HFET device, then both the 2DHG sheet resistance and the contact resistance would saturate, indicating an ohmic contact. An ohmic contact has little to no barrier such that charge transport is unaffected by any additional band bending at the metal/semiconductor interface. However, a Schottky barrier would be affected by further band bending. Either image-force lowering of the barrier height occurs or the barrier thickness is reduced. In either case, the result is an increase in the supply of carriers from an effectively reduced contact resistance [45].

To verify the behavior of the contact resistance past saturation, contact resistance was determined using a second method. The second method used involves subtracting off the sheet resistance component from the total resistance for each resistance measurement and dividing by 2 (for the two current contacts). This was made possible because the sheet resistance was obtained through a four-terminal measurement. The contact resistance obtained by the second method was then averaged across three different devices. The second method data (black) are shown in Figure 5a, where the error bars correspond to the standard deviation. The contact resistance obtained by the first method is overlaid in red. As expected, the two methods for extracting contact resistance were consistent with one another. The Hall density versus gate voltage data in Figure 5b show that a saturation density of 8×10^{11} cm^{-2} is obtained at a gate voltage of −15 V. The mobility versus density data in Figure 5c show that a peak mobility of 2.1×10^5 cm^2 V^{-1} s^{-1}is achieved at a density of 8×10^{11} cm^{-2}. The transport data using Mn_5Ge_3-based contact are consistent with what was previously observed in the same Ge/SiGe heterostructures with PtSiGe-based contacts [13]. The saturation of Hall density with a higher gate voltage is characteristic of Ge/SiGe quantum well systems and is consistent with the saturation of the 2DHG sheet resistance. This further supports that the reduction in contact resistance is due to the capacitive coupling of the gate to the Schottky barrier.

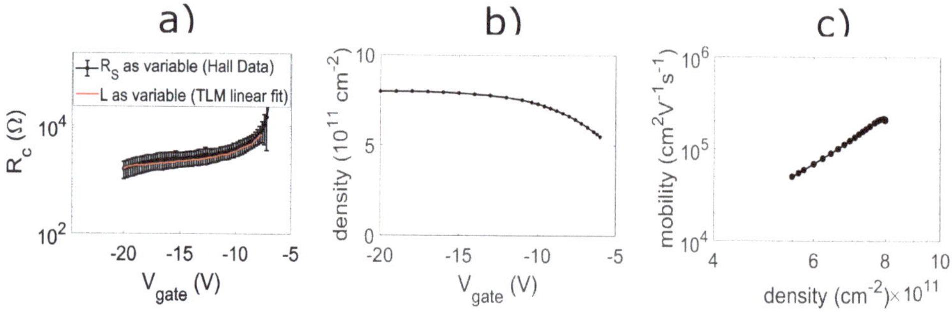

Figure 5. Transport characterization with Mn_5Ge_3 contacts. (**a**) Comparison of contact resistance obtained via Hall data (black) and TLM (red). (**b**) Hall density vs. gate voltage. (**c**) Mobility vs. density.

4. Conclusions

We investigated Mn_5Ge_3-based contacts created via solid-state synthesis on a shallow Ge/SiGe quantum well heterostructure. XRD was used to determine the proper annealing temperature for the phase formation, while STEM HAADF and its FFT image were used to verify the formation of the target Mn-Ge phase. Although our solid-state synthesis could be improved to eliminate the unwanted oxides (Mn_2GeO_4 and MnO) observed, the Mn_5Ge_3-based contacts were observed to work well as p-type contacts for Ge/SiGe HFETs even at cryogenic temperatures. Our electrical characterization results were consistent with what is expected for Mn_5Ge_3 contacts and other p-type contacts on Ge/SiGe.

Our demonstration of the solid-state synthesis of Mn_5Ge_3-based contacts for a shallow Ge/SiGe quantum well heterostructure provides a clear path towards demonstrating spin transport in Ge 2DHG systems similar to what was done in Si 2DEG systems [46] and Ge nanowires [1,2]. The evidence of Schottky barrier lowering, shown from a decreasing contact resistance past saturation, is promising for efficient spin injection and its gate tunability[47]. Future studies should include studying the magnetic properties of $Mn_5(Si_{0.15}Ge_{0.85})_3$ (formed after the removal of the top barrier and quantum well) and spin transport in the Ge/SiGe heterostructure.

Author Contributions: Conceptualization, T.A.H.-D. and T.-M.L.; methodology, T.A.H.-D. and T.-M.L.; software, T.A.H.-D.; validation, T.A.H.-D., S.J.A., P.L. and T.-M.L.; formal analysis, T.A.H.-D. and P.L.; investigation, T.A.H.-D.; resources, T.-M.L.; data curation, T.A.H.-D., S.J.A. and P.L.; writing—original draft preparation, T.A.H.-D.; writing—review and editing, T.A.H.-D., S.J.A., P.L. and T.-M.L.; visualization, T.A.H.-D.; supervision, T.-M.L.; project administration, T.-M.L.; funding acquisition, T.-M.L. All authors have read and agreed to the published version of the manuscript.

Funding: This research received no external funding.

Data Availability Statement: Data will be available upon request.

Acknowledgments: This research was primarily supported by the Laboratory Directed Research and Development Program at Sandia National Laboratories (SNL). This work was performed, in part, at the Center for Integrated Nanotechnologies, an Office of Science user facility operated for the U.S. Department of Energy (DOE) Office of Science. SNL is a multimission laboratory managed and operated by National Technology and Engineering Solutions of Sandia LLC, a wholly owned subsidiary of Honeywell International Inc. for the U.S. DOE National Nuclear Security Administration under contract number DE-NA0003525. This paper describes objective technical results and analysis. Any subjective views or opinions that might be expressed in the paper do not necessarily represent the views of the U.S. DOE or the United States government. The data that support the findings of this study are available from the corresponding author upon reasonable request.

Conflicts of Interest: The authors declare no conflicts of interest.

References

1. Tang, J.; Wang, C.Y.; Hung, M.H.; Jiang, X.; Chang, L.T.; He, L.; Liu, P.H.; Yang, H.J.; Tuan, H.Y.; Chen, L.J.; et al. Ferromagnetic germanide in Ge nanowire transistors for spintronics application. *ACS Nano* **2012**, *6*, 5710–5717. [CrossRef] [PubMed]
2. Tang, J.; Wang, C.Y.; Chang, L.T.; Fan, Y.; Nie, T.; Chan, M.; Jiang, W.; Chen, Y.T.; Yang, H.J.; Tuan, H.Y.; et al. Electrical spin injection and detection in Mn_5Ge_3/Ge/Mn_5Ge_3 nanowire transistors. *Nano Lett.* **2013**, *13*, 4036–4043. [CrossRef] [PubMed]
3. Morrison, C.; Wiśniewski, P.; Rhead, S.; Foronda, J.; Leadley, D.R.; Myronov, M. Observation of Rashba zero-field spin splitting in a strained germanium 2D hole gas. *Appl. Phys. Lett.* **2014**, *105*, 182401. [CrossRef]
4. Hardy, W.J.; Harris, C.T.; Su, Y.H.; Chuang, Y.; Moussa, J.; Maurer, L.N.; Li, J.Y.; Lu, T.M.; Luhman, D.R. Single and double hole quantum dots in strained Ge/SiGe quantum wells. *Nanotechnology* **2019**, *30*, 215202. [CrossRef]
5. Sammak, A.; Sabbagh, D.; Hendrickx, N.W.; Lodari, M.; Paquelet Wuetz, B.; Tosato, A.; Yeoh, L.; Bollani, M.; Virgilio, M.; Schubert, M.A.; et al. Shallow and undoped germanium quantum wells: a playground for spin and hybrid quantum technology. *Adv. Funct. Mater.* **2019**, *29*, 1807613. [CrossRef]
6. Hendrickx, N.; Lawrie, W.; Petit, L.; Sammak, A.; Scappucci, G.; Veldhorst, M. A single-hole spin qubit. *Nat. Commun.* **2020**, *11*, 3478. [CrossRef] [PubMed]
7. Hendrickx, N.; Franke, D.; Sammak, A.; Scappucci, G.; Veldhorst, M. Fast two-qubit logic with holes in germanium. *Nature* **2020**, *577*, 487–491. [CrossRef] [PubMed]
8. Hendrickx, N.W.; Lawrie, W.I.; Russ, M.; van Riggelen, F.; de Snoo, S.L.; Schouten, R.N.; Sammak, A.; Scappucci, G.; Veldhorst, M. A four-qubit germanium quantum processor. *Nature* **2021**, *591*, 580–585. [CrossRef]
9. Van Riggelen, F.; Hendrickx, N.; Lawrie, W.; Russ, M.; Sammak, A.; Scappucci, G.; Veldhorst, M. A two-dimensional array of single-hole quantum dots. *Appl. Phys. Lett.* **2021**, *118*, 044002. [CrossRef]
10. Miller, A.J.; Hardy, W.J.; Luhman, D.R.; Brickson, M.; Baczewski, A.; Liu, C.Y.; Li, J.Y.; Lilly, M.P.; Lu, T.M. Effective out-of-plane g factor in strained-Ge/SiGe quantum dots. *Phys. Rev. B* **2022**, *106*, L121402. [CrossRef]
11. Lodari, M.; Kong, O.; Rendell, M.; Tosato, A.; Sammak, A.; Veldhorst, M.; Hamilton, A.; Scappucci, G. Lightly strained germanium quantum wells with hole mobility exceeding one million. *Appl. Phys. Lett.* **2022**, *120*, 122104. [CrossRef]
12. Kong, Z.; Li, Z.; Cao, G.; Su, J.; Zhang, Y.; Liu, J.; Liu, J.; Ren, Y.; Li, H.; Wei, L.; et al. Undoped Strained Ge Quantum Well with Ultrahigh Mobility of Two Million. *ACS Appl. Mater. Interfaces* **2023**, *15*, 28799–28805. [CrossRef] [PubMed]
13. Hutchins-Delgado, T.A.; Miller, A.J.; Scott, R.; Lu, P.; Luhman, D.R.; Lu, T.M. Characterization of Shallow, Undoped Ge/SiGe Quantum Wells Commercially Grown on 8-in. (100) Si Wafers. *ACS Appl. Electron. Mater.* **2022**, *4*, 4482–4489. [CrossRef]
14. Chou, C.T.; Jacobson, N.T.; Moussa, J.E.; Baczewski, A.D.; Chuang, Y.; Liu, C.Y.; Li, J.Y.; Lu, T.M. Weak anti-localization of two-dimensional holes in germanium beyond the diffusive regime. *Nanoscale* **2018**, *10*, 20559–20564. [CrossRef]
15. Virgilio, M.; Grosso, G. Optical spin orientation in strained Ge/SiGe quantum wells: A tight-binding approach. *Phys. Rev. B* **2009**, *80*, 205309. [CrossRef]
16. Pezzoli, F.; Bottegoni, F.; Trivedi, D.; Ciccacci, F.; Giorgioni, A.; Li, P.; Cecchi, S.; Grilli, E.; Song, Y.; Guzzi, M.; et al. Optical spin injection and spin lifetime in Ge heterostructures. *Phys. Rev. Lett.* **2012**, *108*, 156603. [CrossRef]
17. Schmidt, G.; Ferrand, D.; Molenkamp, L.; Filip, A.; Van Wees, B. Fundamental obstacle for electrical spin injection from a ferromagnetic metal into a diffusive semiconductor. *Phys. Rev. B* **2000**, *62*, R4790. [CrossRef]
18. Fert, A.; Jaffres, H. Conditions for efficient spin injection from a ferromagnetic metal into a semiconductor. *Phys. Rev. B* **2001**, *64*, 184420. [CrossRef]
19. Jeon, K.R.; Min, B.C.; Jo, Y.H.; Lee, H.S.; Shin, I.J.; Park, C.Y.; Park, S.Y.; Shin, S.C. Electrical spin injection and accumulation in CoFe/MgO/Ge contacts at room temperature. *Phys. Rev. B* **2011**, *84*, 165315. [CrossRef]
20. Zhou, Y.; Han, W.; Chang, L.T.; Xiu, F.; Wang, M.; Oehme, M.; Fischer, I.A.; Schulze, J.; Kawakami, R.K.; Wang, K.L. Electrical spin injection and transport in germanium. *Phys. Rev. B* **2011**, *84*, 125323. [CrossRef]
21. Iba, S.; Saito, H.; Spiesser, A.; Watanabe, S.; Jansen, R.; Yuasa, S.; Ando, K. Spin accumulation and spin lifetime in p-type germanium at room temperature. *Appl. Phys. Express* **2012**, *5*, 053004. [CrossRef]
22. Jain, A.; Vergnaud, C.; Peiro, J.; Le Breton, J.; Prestat, E.; Louahadj, L.; Portemont, C.; Ducruet, C.; Baltz, V.; Marty, A.; et al. Electrical and thermal spin accumulation in germanium. *Appl. Phys. Lett.* **2012**, *101*, 022402. [CrossRef]
23. Sharma, S.; Spiesser, A.; Dash, S.P.; Iba, S.; Watanabe, S.; Van Wees, B.; Saito, H.; Yuasa, S.; Jansen, R. Anomalous scaling of spin accumulation in ferromagnetic tunnel devices with silicon and germanium. *Phys. Rev. B* **2014**, *89*, 075301. [CrossRef]
24. Picozzi, S.; Continenza, A.; Freeman, A. First-principles characterization of ferromagnetic Mn_5Ge_3 for spintronic applications. *Phys. Rev. B* **2004**, *70*, 235205. [CrossRef]
25. Spiesser, A.; Saito, H.; Jansen, R.; Yuasa, S.; Ando, K. Large spin accumulation voltages in epitaxial Mn_5Ge_3 contacts on Ge without an oxide tunnel barrier. *Phys. Rev. B* **2014**, *90*, 205213. [CrossRef]
26. Ponath, P.; Posadas, A.; Demkov, A. Ge (001) surface cleaning methods for device integration. *Appl. Phys. Rev.* **2017**, *4*, 021308. [CrossRef]
27. Wittmer, M.; Nicolet, M.A.; Mayer, J. The first phase to nucleate in planar transition metal-germanium interfaces. *Thin Solid Films* **1977**, *42*, 51–59. [CrossRef]
28. Gokhale, A.; Abbaschian, R. The Ge-Mn (germanium-manganese) system. *J. Phase Equilibria* **1990**, *11*, 460–468. [CrossRef]

29. Gates-Rector, S.; Blanton, T. The powder diffraction file: A quality materials characterization database. *Powder Diffr.* **2019**, *34*, 352–360. [CrossRef]
30. Olive-Mendez, S.; Spiesser, A.; Michez, L.; Le Thanh, V.; Glachant, A.; Derrien, J.; Devillers, T.; Barski, A.; Jamet, M. Epitaxial growth of Mn_5Ge_3/Ge (111) heterostructures for spin injection. *Thin Solid Films* **2008**, *517*, 191–196. [CrossRef]
31. Spiesser, A.; Olive-Mendez, S.; Dau, M.T.; Michez, L.; Watanabe, A.; Le Thanh, V.; Glachant, A.; Derrien, J.; Barski, A.; Jamet, M. Effect of thickness on structural and magnetic properties of Mn_5Ge_3 films grown on Ge (111) by solid phase epitaxy. *Thin Solid Films* **2010**, *518*, S113–S117. [CrossRef]
32. Nishimura, T.; Nakatsuka, O.; Akimoto, S.; Takeuchi, W.; Zaima, S. Crystalline orientation dependence of electrical properties of Mn Germanide/Ge (1 1 1) and (0 0 1) Schottky contacts. *Microelectron. Eng.* **2011**, *88*, 605–609. [CrossRef]
33. Xie, Y.; Yuan, Y.; Wang, M.; Xu, C.; Hübner, R.; Grenzer, J.; Zeng, Y.J.; Helm, M.; Zhou, S.; Prucnal, S. Epitaxial Mn_5Ge_3 (100) layer on Ge (100) substrates obtained by flash lamp annealing. *Appl. Phys. Lett.* **2018**, *113*, 222401. [CrossRef]
34. Abbes, O.; Portavoce, A.; Le Thanh, V.; Girardeaux, C.; Michez, L. Phase formation during Mn thin film reaction with Ge: Self-aligned germanide process for spintronics. *Appl. Phys. Lett.* **2013**, *103*, 172405. [CrossRef]
35. Myagkov, V.; Zhigalov, V.; Matsynin, A.; Bykova, L.; Mikhlin, Y.L.; Bondarenko, G.; Patrin, G.; Yurkin, G.Y. Formation of ferromagnetic germanides by solid-state reactions in 20 Ge/80 Mn films. *Thin Solid Films* **2014**, *552*, 86–91. [CrossRef]
36. Myagkov, V.; Bykova, L.; Matsynin, A.; Volochaev, M.; Zhigalov, V.; Tambasov, I.; Mikhlin, Y.L.; Velikanov, D.; Bondarenko, G. Solid state synthesis of Mn_5Ge_3 in Ge/Ag/Mn trilayers: Structural and magnetic studies. *J. Solid State Chem.* **2017**, *246*, 379–387. [CrossRef]
37. Cowley, J.M. *Diffraction Physics*; Elsevier: Amsterdam, The Netherlands, 1995.
38. Adachi, S. *Properties of Semiconductor Alloys: Group-IV, III-V and II-VI Semiconductors*; John Wiley & Sons: Hoboken, NJ, USA, 2009.
39. Chuang, S.L. *Physics of Photonic Devices*; John Wiley & Sons: Hoboken, NJ, USA, 2012.
40. Mamor, M.; Nur, O.; Karlsteen, M.; Willander, M.; Auret, F. Fermi-level pinning and Schottky barrier heights on epitaxially grown fully strained and partially relaxed n-type Si_{1-x} Ge_x layers. *J. Appl. Phys.* **1999**, *86*, 6890–6894. [CrossRef]
41. Sellai, A.; Mesli, A.; Petit, M.; Le Thanh, V.; Taylor, D.; Henini, M. Barrier height and interface characteristics of Au/Mn_5Ge_3/Ge (1 1 1) Schottky contacts for spin injection. *Semicond. Sci. Technol.* **2012**, *27*, 035014. [CrossRef]
42. Lepselter, M.; Sze, S. SB-IGFET: An insulated-gate field-effect transistor using Schottky barrier contacts for source and drain. *Proc. IEEE* **1968**, *56*, 1400–1402. [CrossRef]
43. Snyder, J.P.; Helms, C.; Nishi, Y. Experimental investigation of a PtSi source and drain field emission transistor. *Appl. Phys. Lett.* **1995**, *67*, 1420–1422. [CrossRef]
44. Li, R.; Yao, H.; Lee, S.; Chi, D.; Yu, M.; Lo, G.; Kwong, D. Metal-germanide Schottky source/drain transistor on germanium substrate for future CMOS technology. *Thin Solid Films* **2006**, *504*, 28–31. [CrossRef]
45. Sze, S.M.; Li, Y.; Ng, K.K. *Physics of Semiconductor Devices*; John Wiley & Sons: Hoboken, NJ, USA, 2021.
46. Chang, L.T.; Fischer, I.A.; Tang, J.; Wang, C.Y.; Yu, G.; Fan, Y.; Murata, K.; Nie, T.; Oehme, M.; Schulze, J.; et al. Electrical detection of spin transport in Si two-dimensional electron gas systems. *Nanotechnology* **2016**, *27*, 365701. [CrossRef] [PubMed]
47. Dankert, A.; Dulal, R.S.; Dash, S.P. Efficient spin injection into silicon and the role of the Schottky barrier. *Sci. Rep.* **2013**, *3*, 3196. [CrossRef] [PubMed]

nanomaterials

Article

Circular Photogalvanic Current in Ni-Doped Cd_3As_2 Films Epitaxied on GaAs(111)B Substrate

Gaoming Liang [1,2], Guihao Zhai [1,2], Jialin Ma [1,2], Hailong Wang [1,2], Jianhua Zhao [1,2], Xiaoguang Wu [1,2] and Xinhui Zhang [1,2,*]

1 State Key Laboratory of Superlattices and Microstructures, Institute of Semiconductors, Chinese Academy of Sciences, Beijing 100083, China; gmliang@semi.ac.cn (G.L.); zhai_guihao@semi.ac.cn (G.Z.); jlma1991@semi.ac.cn (J.M.); allen@semi.ac.cn (H.W.); jhzhao@red.semi.ac.cn (J.Z.); xgwu@red.semi.ac.cn (X.W.)

2 Center of Materials Science and Optoelectronics Engineering, University of Chinese Academy of Sciences, Beijing 100049, China

* Correspondence: xinhuiz@semi.ac.cn

Abstract: Magnetic element doped Cd_3As_2 Dirac semimetal has attracted great attention for revealing the novel quantum phenomena and infrared opto-electronic applications. In this work, the circular photogalvanic effect (CPGE) was investigated at various temperatures for the Ni-doped Cd_3As_2 films which were grown on GaAs(111)B substrate by molecular beam epitaxy. The CPGE current generation was found to originate from the structural symmetry breaking induced by the lattice strain and magnetic doping in the Ni-doped Cd_3As_2 films, similar to that in the undoped ones. However, the CPGE current generated in the Ni-doped Cd_3As_2 films was approximately two orders of magnitude smaller than that in the undoped one under the same experimental conditions and exhibited a complex temperature variation. While the CPGE current in the undoped film showed a general increase with rising temperature. The greatly reduced CPGE current generation efficiency and its complex variation with temperature in the Ni-doped Cd_3As_2 films was discussed to result from the efficient capture of photo-generated carriers by the deep-level magnetic impurity bands and enhanced momentum relaxation caused by additional strong impurity scattering when magnetic dopants were introduced.

Keywords: Ni-doped Dirac semimetal Cd_3As_2; the circular photogalvanic effect; the deep-level magnetic impurity bands

Citation: Liang, G.; Zhai, G.; Ma, J.; Wang, H.; Zhao, J.; Wu, X.; Zhang, X. Circular Photogalvanic Current in Ni-Doped Cd_3As_2 Films Epitaxied on GaAs(111)B Substrate. *Nanomaterials* **2023**, *13*, 1979. https://doi.org/10.3390/nano13131979

Academic Editor: Julian Maria Gonzalez Estevez

Received: 25 May 2023
Revised: 26 June 2023
Accepted: 28 June 2023
Published: 29 June 2023

1. Introduction

As a representative material of the three-dimensional Dirac semimetals (3D DSMs), Cd_3As_2 has been extensively studied owing to its outstanding features of good stability in air, high carrier mobility and Fermi velocity [1–3]. Its topological nature has been verified by the angle-resolved photoemission spectroscopy (ARPES) and quantum transport experiments [2–6]. The giant linear magnetoresistance [1,7,8], Landau levels [5,9], quantum Hall effect [10,11], superconductivity [12,13] and magneto-optical phenomena [14], etc., have been studied to reveal its novel physical properties. Cd_3As_2 is also proposed to be suitable for high speed, broadband photodetectors, or mid-infrared optical switches and modulators due to its advantages of having high carrier mobility and linear dispersion of band energy [15–20].

Circular photogalvanic effect (CPGE) can generate a spin-polarized charge current by creating an asymmetric carrier population in momentum space [21,22], under circularly-polarized light excitation with the angular momentum selection rule. CPGE has served as an excellent experimental method to study the spin-associated properties in low-dimensional semiconductors [23–26], topological insulators (TIs) [27–31], Weyl semimetals [32–39], and 2D Dirac material graphene [40–44]. In theory, the CPGE was predicted to be zero in an ideal Dirac system because of the symmetric photoexcitation with respect to the Dirac point [39,45,46], as a consequence of both the time-reversal and spatial inversion symmetry in DSMs. However, under symmetry breaking caused by the epitaxial strain or Schottky electric field, CPGE has been observed experimentally in Cd_3As_2 films and Cd_3As_2 Nanobelts [47,48]. It is revealed that symmetry breaking in DSMs can be induced by introducing strain, dimensional control, or doping magnetic elements [5,49–51]. The electronic and magnetic properties of 3d transition metal element (3d-TM) doped Cd_3As_2 have been theoretically investigated by using first principle calculation with density functional theory. Cr-doped Cd_3As_2 was proposed to be a promising room-temperature ferromagnetic semiconductor via Monte Carlo simulations [52]. It is reported that 3d-TM doping reduces the symmetry of Cd_3As_2 in the Cr- and Mn-doped Cd_3As_2, leading to a band gap opening [15,16,52–55]. Meanwhile, magnetic doping was found to make a great difference in the relaxation time of photoexcited carriers by ultrafast spectral experiments in the Cr- and Mn-doped Cd_3As_2 [15,56]. In addition, Mn doping was reported to influence the transport properties of Cd_3As_2 thin films grown on GaAs(111)B substrate [57]. Given that magnetic doping could cause symmetry breaking and possible phase transition from DSM to Weyl semimetal for Cd_3As_2, it is highly desirable to explore CPGE photocurrent in magnetic element doped Cd_3As_2, to further explore its spin-related physical properties and potential opto-electronic applications. However, there is no CPGE study available yet in magnetically doped Cd_3As_2 so far.

In this work, we systematically measured the CPGE current generation for the Ni-doped Cd_3As_2 films with different doping concentrations and compared it with that in the undoped one at various temperatures. The CPGE current can be observed in all Cd_3As_2 films and shows a general increase with a larger optical incident angle and pumping power. The CPGE current measured in the Ni-doped Cd_3As_2 films was greatly reduced compared with that from an undoped one under the same experimental conditions. The CPGE current generated in the Ni-doped Cd_3As_2 films was suggested to originate from the reduced structural symmetry by the large epitaxial strain and magnetic doping, rather than a possible phase transition to Weyl semimetal by magnetic doping. Moreover, a complex temperature variation of CPGE current was observed in the Ni-doped Cd_3As_2 films, which was attributed to the efficient capture of photo-generated carriers upon optical excitation by the deep-level magnetic impurity bands and the enhanced momentum relaxation with magnetic element doping.

2. Materials and Methods

The experiment was carried out on the 20 nm-thick Ni-doped Cd_3As_2 films with different doping concentrations and the undoped control sample. All films were grown on GaAs(111)B substrate by molecular beam epitaxy at a low temperature of 180 °C. During the growth process, the molecular beams of Cd, Ni and As were controlled by using three isolated effusion cells. For the Ni-doped Cd_3As_2 films with different doping concentrations, the beam equivalent pressure ratio of Ni to Cd was set to be 2%, 4% and 8%, and the corresponding Ni source growth temperature was 1130 °C, 1170 °C and 1210 °C, respectively. The nominal Ni concentration was 2%, 4% and 8% and denoted as sample-A (Ni-2%), sample-B (Ni-4%) and sample-C (Ni-8%), respectively. Since the epitaxial Cd_3As_2 thin films would be easily oxidized if exposed to the air, and this can cause an adverse effect of electrical and optical properties, a 3-nm-thick GaAs capping layer was grown at the same low temperature of 180 °C to prevent the film from oxidation. The Curie temperature of the Ni-doped Cd_3As_2 films is about 45 K. More detailed sample growth procedures and characterization results can be found in our previous studies [47,58]. The experimental setup is shown schematically in Figure 1: the x-y coordinates plane is defined as being parallel to the sample's surface and the z-axis is perpendicular to the sample's surface. We made a pair of ohmic electrodes by indium deposition along the x-axis with a 3 mm distance far apart. A Ti: sapphire laser (Chameleon Ultra II, Coherent Inc., Santa Clara, CA, USA) with a repetition rate of 80 MHz, was utilized as an exciting light source in our experiment. The oblique incidence plane of the laser is perpendicular to the x-axis. The irradiated laser spot is tuned to locate in between the two electrodes with a radius of about 1 mm (the cleaved sample is about 5 × 5 mm in size). We used a polarizer and a $\lambda/4$ wave-plate to alter the light helicity $Pc = \sin 2\varphi$ from the left-handed (σ^-, $Pc = -1$) to right-handed (σ^+, $Pc = 1$), here φ is the angle between the polarization direction of the polarizer and the optical axis of the $\lambda/4$ plate. Then the photocurrent is measured by a preamplifier and a lock-in amplifier that is in phase with an optical chopper working at a fixed frequency. All samples were placed in a Janis closed-cycle optical cryostat for low-temperature measurements.

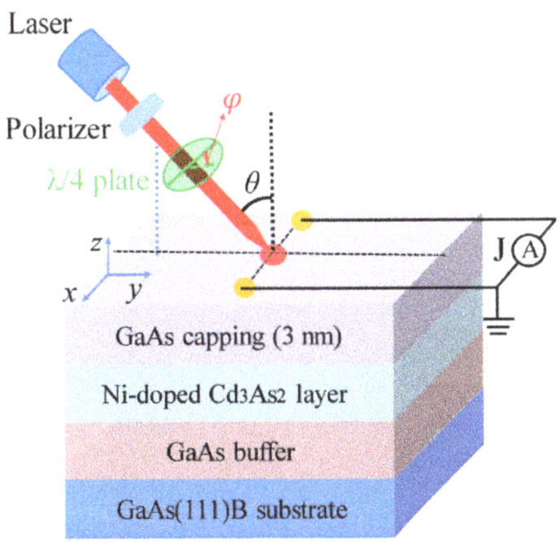

Figure 1. The experimental sketch showing the photo excitation and current measurement. The optical incident angle is θ. The x-y coordinates plane is defined to be parallel to the sample's surface. The red circle in the center is the laser spot, and the yellow circles along the x-axis are the ohmic electrodes. A $\lambda/4$ wave plate is utilized to tune the light helicity (by tuning the angle φ between the linear polarization direction of the excitation laser and the fast axis of the $\lambda/4$ plate). The photocurrent is collected along the x direction.

3. Results and Discussion

Figure 2a,b shows the representative photogalvanic current response by varying angle φ measured at 10 K and an incident angle of 45°, under excitation of 750 nm and 900 nm with pumping power of 14 mW for the Ni-4% doped sample-B. Though Cd_3As_2 thin films were epitaxied on GaAs substrate and capped with a 3-nm-thick GaAs which was also grown at a low temperature of 180 °C (LT-GaAs) layer, as depicted in the sample preparation part, the possible photogalvanic current contribution from either GaAs buffer/substrate or GaAs capping layers can be ruled out, by considering the facts of nearly amorphous LT-GaAs capping layer and the control study results of a bare GaAs buffer sample, as already discussed in our previous work [47]. In addition, our photogalvanic current study on Ni-4% samples with different excitation wavelengths further confirms this and will be discussed later. The photogalvanic current includes two components with different periodicities and is usually written as [33]

$$j_\lambda = j_C \sin(2\varphi + \psi_C) + j_L \sin(4\varphi + \psi_L) + j_0, \quad (1)$$

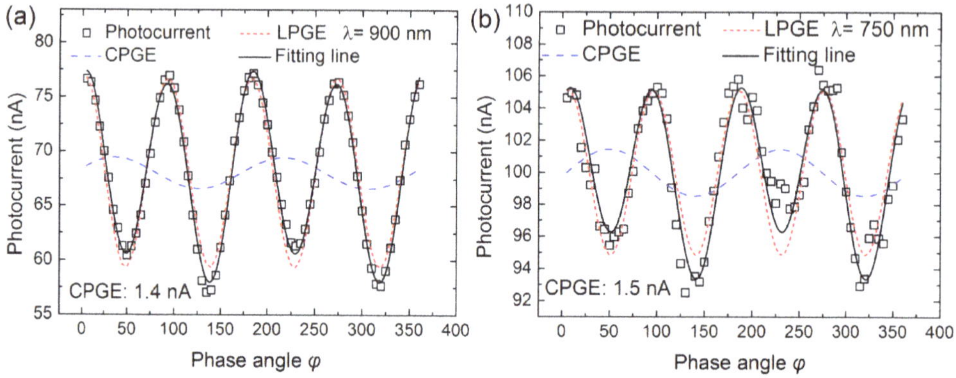

Figure 2. The photocurrent response measured at 10 K by varying phase angle φ for the Ni-4% sample. The incident angle is 45° and the excitation power is 14 mW. The excitation wavelength in (**a**,**b**) is 900 nm and 750 nm, respectively. The open squares represent the raw experimental data, and the black lines represent the fitting curve based on Equation (1). The extracted CPGE and LPGE currents are represented by the blue and red dashed lines, respectively.

Here, j_C and j_L are the amplitude of CPGE and linear photogalvanic effect (LPGE) current, respectively. ψ_C (ψ_L) represents the initial phase of CPGE (LPGE). j_0 is the background current resulting from the photovoltaic effect at electrodes or the Dember effect and does not rely on the helicity of the excitation laser [21]. By fitting the measured photogalvanic current with Equation (1), the CPGE (LPGE) current response can be separately determined and was shown by the blue (red) dashed lines in Figure 2. It was observed that the extracted CPGE current in the Ni-4% sample at the radiation wavelength of 900 nm and 750 nm is 1.4 nA and 1.5 nA, respectively. This finding further suggests that the measured CPGE photocurrent should be generated mainly in the Ni-doped Cd_3As_2 film itself, rather than the upper GaAs capping layer or the GaAs buffer/substrate. Since the photon energy of radiation wavelength at 900 nm is lower than that of GaAs bandgap, thus we can exclude the possible CPGE generation from GaAs capping/buffer layer induced by interband excitation. In addition, the generated CPGE with radiation wavelengths of 900 nm and 750 nm is almost consistent within the experimental error, which further rules out the possible intraband transition associated with CPGE generation in GaAs capping/buffer layer.

Figure 3a shows the extracted CPGE amplitude by fitting with Equation (1) at various incident angles θ in the Ni-4% sample, measured with the excitation power of 21 mW at 900 nm. One can see that, the CPGE current increases with the incident angle within the measured angle range from 0 to 60 degrees, this is consistent with the phenomenological theory of CPGE described by the following formula [21]:

$$j_\lambda = \gamma_{\lambda\mu} i(\mathbf{E} \times \mathbf{E}^*)_\mu, \tag{2}$$

Here, j_λ represents the generated CPGE photocurrent density, $\gamma_{\lambda\mu}$ represents the second-rank pseudo-tensor associated with the symmetry of the material, λ goes through all the three Cartesian coordinates x, y, z, $\mu = x, y, z$, $\mathbf{E}$ represents the complex amplitude of the electric field of the incident excitation light, its unit vector $\hat{e}$ points to the direction of light propagation. $\gamma_{\lambda\mu} i(\mathbf{E} \times \mathbf{E}^*)_\mu$ can be transformed into $\gamma P_{circ} E^2 \hat{e}_{||}$, where P_{circ} denotes the circular polarization degree. That is, the CPGE current is proportional to the field projection of the incident light onto the sample surface ($\propto E^2\hat{e}_{||}$), as well as the radiation intensity. Figure 3b shows the measured CPGE current at 10 K in sample B with various photoexcitation power excited at 900 nm and an incident angle of 45°. It is seen that the CPGE increases nearly linearly with photoexcitation power from 3 to 27 mW, which agrees with the theoretical expectation.

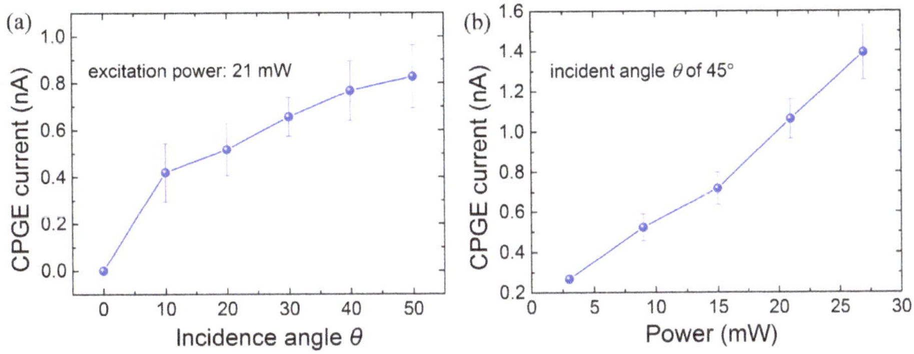

Figure 3. (**a**) The extracted CPGE amplitude at various incident angles θ measured with the excitation power of 21 mW. (**b**) The CPGE photocurrent at various excitation power from 3 to 27 mW measured at the incident angle of 45°. The measurements are done at 10 K with an excitation of 900 nm for the Ni-4% sample-B.

To study the effect of magnetic doping on CPGE current generation, we further measured the helicity-dependent photocurrent responses for the Ni-doped Cd_3As_2 films with different doping concentrations and compared the results with an undoped control sample. The extracted CPGE current by fitting the original data taken at room temperature with Equation (1), normalized with the corresponding photoexcitation power, is shown in Figure 4. The inset shows the zoom-in CPGE results for Cd_3As_2 films with different Ni dopant concentrations. We can see that the measured CPGE current in the undoped 20 nm Cd_3As_2 is two orders of magnitude larger than that in the Ni-doped Cd_3As_2 films. For the Ni-doped Cd_3As_2 thin films, the magnetic element doping can introduce time inversion symmetry breaking as theoretically expected, leading to possible phase transition from Dirac to Weyl semimetals, so that stronger CPGE current than the undoped ones might be expected. However, our experiment shows that the magnetic element doping did not enhance the CPGE current generation, on the contrary, the generated CPGE current in the Ni-doped Cd_3As_2 films is much smaller than that in the undoped one, as shown in Figure 4. Therefore, we believe that the CPGE current generation in the Ni-doped Cd_3As_2 films also originates from the structural symmetry breaking induced by lattice strain and Ni doping, similar to that in the undoped Cd_3As_2 thin films discussed in our earlier study [47]. It would be hard to give a quantitative estimation of the additional strain introduced by 2%, 4%, and 8% Ni doping, though, the dominant strain in the Ni-doped Cd_3As_2 films is believed to result from the large lattice mismatch of 10% between Cd_3As_2 and GaAs(111)B substrate, the same as the undoped ones. To analyze the observed difference in the CPGE current generation efficiency in the doped and undoped films, we first compared the parameters (Electron density and Hall mobility) closely related to the CPGE current response, thanks to the characterization results of our samples in previous work [58]. Table 1 lists the electron density and Hall mobility at 3 K for the Ni-doped Cd_3As_2 films with different doping concentrations and the undoped ones. It is seen that the electron density and Hall mobility of the Ni-doped Cd_3As_2 films with different doping concentration and the undoped one is almost on the same order of magnitude, which cannot account for the huge difference in CPGE current generation as shown in Figure 4. We thus further compare the photo-excited carrier generation and its relaxation dynamics for the doped and undoped films, since CPGE current is also closely associated with the momentum relaxation process and carrier lifetime.

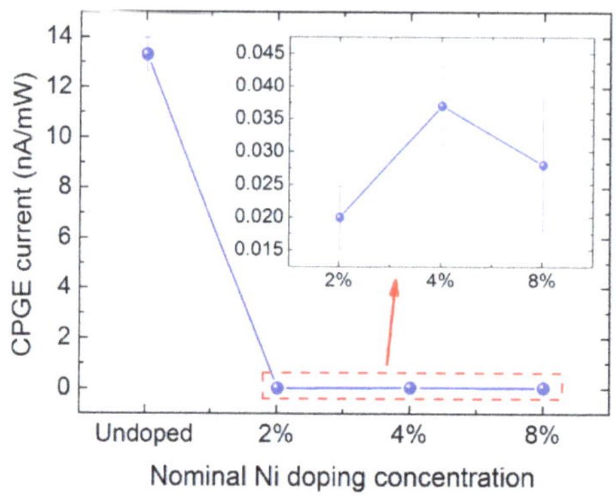

Figure 4. The measured CPGE current normalized with the corresponding photoexcitation power in the Ni-doped Cd_3As_2 films with different Ni doping concentrations. The measurements are done at room temperature with excitation of 900 nm at an incident angle of 45°. The inset shows the zoom-in results for the doped films of Ni-2%, Ni-4% and Ni-8%, respectively.

Table 1. The electron density and Hall mobility at 3 K for different samples [58].

Sample	Undoped	Sample-A	Sample-B	Sample-C
Eectron density (cm^{-2})	7.8×10^{11}	8.0×10^{11}	1.4×10^{12}	1.1×10^{12}
Hall mobility ($cm^2 V^{-1} s^{-1}$)	710	1000	560	260

Figure 5 shows typical transient reflection spectra ($\Delta R/R$) of different samples, measured with pump and probe wavelengths of 800 nm and 3500 nm at 4.5 K, and more transient dynamics results can be found in our previous work [59]. Apparently, the photo-excited carrier decay time of the Ni-doped films, take the result of sample-B as an example, is only ~0.75 ps, much shorter than the undoped film (~3.98 ps). This shorter photo-excited carrier lifetime in the Ni-doped films results from the greatly enhanced electron-phonon and electron-impurity scattering introduced by Ni doping [56,59]. Meanwhile, the absolute peak value of $\Delta R/R$ for the undoped sample was found to be about three times of the doped ones. The peak value of $\Delta R/R$ is closely relevant to the optically excited carrier concentration according to previous work [60]. As a result, compared with the undoped Cd_3As_2 film, the Ni-doped samples have a shorter carrier lifetime and lower photo-excited carrier density under the same excitation condition. In addition, our previous study also showed that there appeared a long decay time of tens of picoseconds at the elevated temperature or higher optical pumping power, for all Ni-doped films [59]. This long carrier decay process has also been reported in Mn-doped Cd_3As_2 and was associated with the deep-level magnetic impurity level near the Cd_3As_2 Fermi surface [56]. The rather weak temperature variation of this long decay process is a characteristic of the trapped carriers by the magnetic impurity states in Mn- or Ni-doped Cd_3As_2 films [56,59]. Our transient optical reflection studies for the measured films, together with their electrical characterization, suggest that the efficient capture of photo-generated carriers by the deep-level magnetic impurity bands, and the greatly enhanced momentum relaxation caused by stronger electron-phonon and electron-impurity scattering introduced by Ni doping, effectively reduce the CPGE current generation in the Ni-doped Cd_3As_2 films compared with the undoped one.

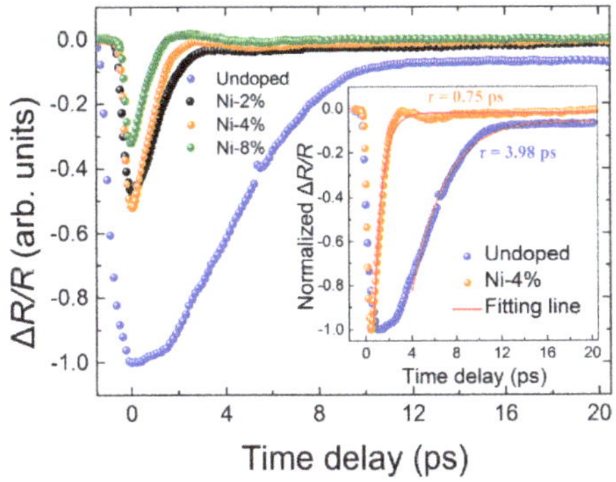

Figure 5. The $\Delta R/R$ responses of Cd_3As_2 films with different Ni doping concentrations at 4.5 K, measured with the pump and probe wavelength of 800 nm and 3500 nm, respectively. All the $\Delta R/R$ responses were normalized relative to the minimum value of the undoped sample. The inset shows the fitting results for the undoped and doped (sample B, Ni-4%) samples, here the $\Delta R/R$ responses were normalized relative to their own minimum values of each sample.

Furthermore, we made a comparison investigation on the helicity-dependent photocurrent for the undoped Cd_3As_2 film and sample-B (Ni-4%) at various temperatures from 10 to 300 K. Figure 6a shows the extracted CPGE current as a function of temperature in sample-B excited at 900 nm and the undoped Cd_3As_2 film excited at 750 nm, here the generated CPGE current is normalized by the optical excitation power. It is seen that the CPGE current generation efficiency in the undoped Cd_3As_2 film shows a general increase with rising temperature. However, for the Ni-4% doped sample, the CPGE current generation efficiency shows a complex temperature variation: it first gradually increases when increasing temperature from 10 to 75 K, but then decreases in the temperature range of 75–200 K, and eventually increases again with rising temperature from 200 to 300 K. Since the photon energy at both excitation wavelength of 750 nm and 900 nm is much higher than the linear Dirac band energy range of Cd_3As_2, in addition, the CPGE current generated with excitation wavelength of 750 nm and 900 nm is almost the same under the same experimental conditions as shown in Figure 2, so the different temperature variation of the CPGE current generation efficiency excited at 750 nm and 900 nm for the undoped and Ni-doped Cd_3As_2 films, should be mainly caused by Ni doping, rather than the electron band structure. Therefore, we further discuss the effect of Ni doping on CPGE current in Cd_3As_2 films. According to previous works [61,62], the CPGE current can be described by

$$J_{CPGE} = e\alpha_e G\tau_p P_c S\mu, \tag{3}$$

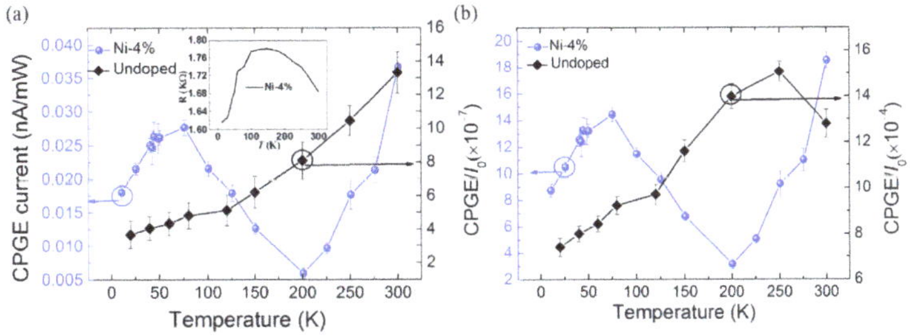

Figure 6. (**a**) The extracted CPGE generation efficiency at various temperatures, obtained by normalization of the measured photocurrent with respect to the corresponding pumping power,

for the doped film of Ni-4% (the blue dots) and the undoped one (the black diamond), respectively. The measurements are done at an incident angle of 45° under excitation of 900 nm and 750 nm, respectively. The inset shows the temperature-dependent resistance measured for the doped film of Ni-4%. (**b**) Temperature dependence of CPGE current normalized relative to the photoconductive current I_0 for the doped film of Ni-4% (the blue dots) and the undoped one (the black diamond), respectively.

Here, α_e represents the effective electric field caused by the spin-orbit coupling (SOC), e is the elementary charge, G is the generation rate of the spin-polarized electrons, τ_p and μ are the momentum relaxation time and mobility of the electrons, P_c is the circular polarization degree, S is the cross-section area of the current. Under an applied DC bias, the photoconductivity current I_0 can be written as [61,62]

$$I_0 = eEG\tau_0 S\mu, \tag{4}$$

Here, E is the applied electric field and τ_0 is the recombination lifetime of the photo-excited carriers. To eliminate the influence of the optical absorptivity and the electron mobility at various temperatures, one can normalize the CPGE current with respect to I_0 as

$$\frac{J_{CPGE}}{I_0} = \frac{\alpha_e \tau_p}{E\tau_0}, \tag{5}$$

In our experiment, we measured the temperature-dependent resistance of the Ni-4% sample as shown in the inset of Figure 6a. One can see that the resistance first increases with temperature (10–150 K) and then decreases within a temperature range of 150–300 K, indicating that the Ni-doped Cd_3As_2 film undergoes a phase transition from semimetal to semiconductor (a trivial insulator phase). However, the measured resistance of the undoped Cd_3As_2 film decreases gradually with temperature [58]. By assuming a DC bias of 1 V applied to the device, the photoconductivity current I_0 can be roughly estimated by the measured resistance. Figure 6 shows the extracted CPGE current normalized by the excitation power and photoconductivity current I_0, respectively, as a function of temperature for the doped film of Ni-4% and the undoped one.

As can be seen from Figure 6b, after eliminating the influence of the optical absorptivity and the electron mobility, the temperature variation trend of the normalized CPGE current relative to the photoconductivity current I_0 for both the doped and undoped films is almost the same as that in Figure 6a, implying that the optical absorptivity and electron mobility are not the main factors affecting CPGE current variation with temperature. A previous study has shown that α_e increases slowly with temperature [63]. The temperature variable τ_p is generally more significant than that of τ_0. In addition, our previous work has shown that the photoexcited carrier relaxation time (a few picoseconds) in Cd_3As_2 films gradually increases with temperature [59,64]. And the carrier density is known to increase with temperature as well [58,65]. Therefore, under the same excitation pump power, the higher-density carriers will be excited at the elevated temperature, though the electron mobility decreases with increasing temperature. Moreover, it is expected that the enhanced Coulomb screening at higher carrier density would efficiently weaken the electron-electron and electron-phonon scattering [64], thus can slow down the momentum relaxation process with increasing temperature. This factor may be the main reason for the general CPGE current increase with rising temperature in the undoped Cd_3As_2 film observed in our experiment, though we cannot quantitatively describe the exact influence of different factors on the temperature variation of CPGE current.

For Ni-doped Cd_3As_2 thin films, though its Curie temperature T_c was determined to be around 45 K [57,58], there is no obvious abnormal change of CPGE near T_c observed as seen in Figure 6, indicating that magnetism has little influence on CPGE current in the Ni-doped Cd_3As_2 films. In the temperature range of 10–75 K, CPGE current gradually increases with temperature, this tendency is consistent with that in the undoped Cd_3As_2 film shown in Figure 6. We believe that it has the same reason, that is, the increased carrier density and suppressed momentum relaxation with rising temperature lead to the increase of CPGE current. As temperature rises beyond 75 K, the increased Fermi level could be closer or even higher than the deep-level magnetic impurity state introduced by Ni doping, in addition, the occupation probability of states nearer the Fermi energy is strongly affected by the smeared Fermi surface upon optical excitation [66]. These facts can promote the efficient capture of photo-generated carriers by the deep-level magnetic impurity bands. The trapped electrons are unable to contribute effective photocurrent, resulting in an obvious decrease of CPGE current. As a matter of fact, our previous transient reflection studies could reveal the long decay time of tens of picoseconds associated with the trapped carriers for the Ni-doped films only at temperatures above 100 K [59]. This is, to some extent, in line with the drastic decrease of CPGE above 75 K observed here for the Ni-doped film. When temperature increases up to 200 K, it is seen that the CPGE current increases again when temperature increases. The possible reason for this phenomenon may be that the impurity states near the Fermi level gradually ionize at higher temperatures, and more free carriers can contribute to CPGE current generation. So that the overall CPGE current increases gradually again with temperature, similar to that in the undoped Cd_3As_2 film.

4. Conclusions

We systematically investigated the CPGE current generation in the epitaxial Ni-doped Cd_3As_2 films with different doping concentrations at various temperatures. Our experimental results show that, though the Ni-doped Cd_3As_2 films become ferromagnetic below 45 K owing to the magnetic doping, the observed CPGE current in the Ni-doped Cd_3As_2 film does not get enhanced by the possible transition from DSM to Weyl semimetal owing to the broken time-reversal symmetry. Instead, we observed two orders of magnitude smaller CPGE current in the Ni-doped Cd_3As_2 films than that in the undoped one under the same experimental conditions. The CPGE generation in the Ni-doped Cd_3As_2 films was attributed to the reduced structural symmetry by the large epitaxial strain and magnetic doping, the same as the undoped film. Moreover, the CPGE current in the Ni-doped Cd_3As_2 films exhibits a complex change when varying temperature, unlike the undoped film which shows a generally increased CPGE current with rising temperature. The greatly reduced CPGE current generation efficiency in the Ni-doped Cd_3As_2 films, together with its complex temperature variation, was closely connected with the efficient capture of photo-generated carriers by the deep-level magnetic impurity bands and the enhanced momentum relaxation caused by strong impurity scattering when doped with magnetic elements. An improved magnetic doping technique with reduced or absent magnetic impurity levels is expected to greatly enhance the CPGE current generation efficiency in magnetically doped Cd_3As_2 films.

Author Contributions: Conceptualization, guidance, data analysis, X.Z.; Experiments and data analysis, G.L., G.Z., X.W.; Writing—original draft, G.L.; Samples growth and characterization, J.M., H.W., J.Z.; Writing—review & editing, X.Z., X.W., H.W. All authors have read and agreed to the published version of the manuscript.

Funding: This research was funded by the National Key Research and Development Program of China (No. 2022YFA1405100), the National Natural Science Foundation of China (No. 11774337 and 12174383), and the Strategic Priority Research Program of the Chinese Academy of Sciences (No. XDB43000000). H. L. Wang also acknowledges the support from the Youth Innovation Promotion Association, Chinese Academy of Sciences (No. 2021110).

Data Availability Statement: The data supporting the findings of this study are available from the corresponding author upon reasonable request.

Conflicts of Interest: The authors declare no competing financial interests.

References

1. Liang, T.; Gibson, Q.; Ali, M.N.; Liu, M.; Cava, R.J.; Ong, N.P. Ultrahigh mobility and giant magnetoresistance in the Dirac semimetal Cd_3As_2. *Nat. Mater.* **2015**, *14*, 280. [CrossRef]
2. Liu, Z.K.; Jiang, J.; Zhou, B.; Wang, Z.J.; Zhang, Y.; Weng, H.M.; Prabhakaran, D.; Mo, S.-K.; Peng, H.; Dudin, P.; et al. A stable three-dimensional topological Dirac semimetal Cd_3As_2. *Nat. Mater.* **2014**, *13*, 677. [CrossRef]
3. Borisenko, S.; Gibson, Q.; Evtushinsky, D.; Zabolotnyy, V.; Buchner, B.; Cava, R.J. Experimental Realization of a Three-Dimensional Dirac Semimetal. *Phys. Rev. Lett.* **2014**, *113*, 027603. [CrossRef]
4. Neupane, M.; Xu, S.Y.; Sankar, R.; Alidoust, N.; Bian, G.; Liu, C.; Belopolski, I.; Chang, T.R.; Jeng, H.T.; Lin, H. Observation of a three-dimensional topological Dirac semimetal phase in high-mobility Cd_3As_2. *Nat. Commun.* **2014**, *5*, 3786. [CrossRef]
5. Wang, Z.; Weng, H.; Wu, Q.; Dai, X.; Fang, Z. Three-dimensional Dirac semimetal and quantum transport in Cd_3As_2. *Phys. Rev. B* **2013**, *88*, 125427. [CrossRef]
6. Zhang, C.; Narayan, A.; Lu, S.; Zhang, J.; Zhang, H.; Ni, Z.; Yuan, X.; Liu, Y.; Park, J.H.; Zhang, E.; et al. Evolution of Weyl orbit and quantum Hall effect in Dirac semimetal Cd_3As_2. *Nat. Commun.* **2017**, *8*, 1272. [CrossRef]
7. Li, C.Z.; Wang, L.X.; Liu, H.; Wang, J.; Liao, Z.M.; Yu, D.P. Giant negative magnetoresistance induced by the chiral anomaly in individual Cd_3As_2 nanowires. *Nat. Commun.* **2015**, *6*, 10137. [CrossRef] [PubMed]
8. Narayanan, A.; Watson, M.D.; Blake, S.F.; Bruyant, N.; Drigo, L.; Chen, Y.L.; Prabhakaran, D.; Yan, B.; Felser, C.; Kong, T.; et al. Linear magnetoresistance caused by mobility fluctuations in the n-doped Cd_3As_2. *Phys. Rev. Lett.* **2015**, *114*, 117201. [CrossRef] [PubMed]
9. Galletti, L.; Schumann, T.; Shoron, O.F.; Goyal, M.; Kealhofer, D.A.; Kim, H.; Stemmer, S. Two-dimensional Dirac fermions in thin films of Cd_3As_2. *Phys. Rev. B* **2018**, *97*, 115132. [CrossRef]
10. Goyal, M.; Galletti, L.; Salmani-Rezaie, S.; Schumann, T.; Kealhofer, D.A.; Stemmer, S. Thickness dependence of the quantum Hall effect in films of the three-dimensional Dirac semimetal Cd_3As_2. *APL Mater.* **2018**, *6*, 026105. [CrossRef]
11. Uchida, M.; Nakazawa, Y.; Nishihaya, S.; Akiba, K.; Kriener, M.; Kozuka, Y.; Miyake, A.; Taguchi, Y.; Tokunaga, M.; Nagaosa, N.; et al. Quantum Hall states observed in thin films of Dirac semimetal Cd_3As_2. *Nat. Commun.* **2017**, *8*, 2274. [CrossRef] [PubMed]
12. Suslov, A.V.; Davydov, A.B.; Oveshnikov, L.N.; Morgun, L.A.; Kugel, K.I.; Zakhvalinskii, V.S.; Pilyuk, E.A.; Kochura, A.V.; Kuzmenko, A.P.; Pudalov, V.M.; et al. Observation of subkelvin superconductivity in Cd_3As_2 thin films. *Phys. Rev. B* **2019**, *99*, 094512. [CrossRef]
13. Wang, H.; Wang, H.C.; Liu, H.; Hong, L.; Yang, W.; Shuang, J.; Liu, X.J.; Xie, X.C.; Wei, J.; Wang, J. Observation of superconductivity induced by a point contact on 3D Dirac semimetal Cd_3As_2 crystals. *Nat. Mater.* **2016**, *15*, 38. [CrossRef] [PubMed]
14. Akrap, A.; Hakl, M.; Tchoumakov, S.; Crassee, I.; Kuba, J.; Goerbig, M.O.; Homes, C.C.; Caha, O.; Nov'ak, J.; Teppe, F.; et al. Magneto-optical signature of massless Kane electrons in Cd_3As_2. *Phys. Rev. Lett.* **2016**, *117*, 136401. [CrossRef]
15. Zhu, C.; Wang, F.; Meng, Y.; Xiang, Y.; Xiu, F.; Luo, H.; Wang, Y.; Li, J.; Lv, X.; Liang, H.; et al. A robust and tuneable mid-infrared optical switch enabled by bulk Dirac fermions. *Nat. Commun.* **2017**, *8*, 14111. [CrossRef] [PubMed]
16. Sun, Y.; Meng, Y.; Jiang, H.; Qin, S.; Yang, Y.; Xiu, F.; Shi, Y.; Zhu, S.; Wang, F. Dirac semimetal saturable absorber with actively tunable modulation depth. *Opt. Lett.* **2019**, *44*, 582. [CrossRef]
17. Yang, M.; Wang, J.; Han, J.; Ling, J.; Ji, C.; Kong, X.; Liu, X.; Huang, Z.; Gou, J.; Liu, Z.; et al. Enhanced Performance of Wideband Room Temperature Photodetector Based on Cd_3As_2 Thin Film/Pentacene Heterojunction. *ACS Photon.* **2018**, *5*, 3438. [CrossRef]
18. Dai, Z.J.; Manjappa, M.; Yang, Y.K.; Tan, T.C.W.; Qiang, B.; Han, S.; Wong, L.J.; Xiu, F.X.; Liu, W.; Singh, R. High Mobility 3D Dirac Semimetal (Cd_3As_2) for Ultrafast Photoactive Terahertz Photonics. *Adv. Funct. Mater.* **2021**, *17*, 2011011. [CrossRef]
19. Yavarishad, N.; Hosseini, T.; Kheirandish, E.; Weber, C.P.; Kouklin, N. Room-temperature self-powered energy photodetector based on optically induced Seebeck effect in Cd_3As_2. *Appl. Phys. Express* **2017**, *10*, 052201. [CrossRef]
20. Chorsi, H.T.; Yue, S.; Iyer, P.P.; Goyal, M.; Schumann, T.; Stemmer, S.; Liao, B.; Schuller, J.A. Widely Tunable Optical and Thermal Properties of Dirac Semimetal Cd_3As_2. *Adv. Opt. Mater.* **2020**, *8*, 1901192. [CrossRef]
21. Ganichev, S.D.; Prettl, W. Spin photocurrents in quantum wells. *J. Phys. Condens. Matter* **2003**, *15*, 935. [CrossRef]
22. Ivchenko, E.L. Circular Photo-Galvanic and Spin-Galvanic Effects. *Lect. Notes Phys.* **2005**, *658*, 23–50.
23. Ganichev, S.D.; Ivchenko, E.L.; Danilov, S.N.; Eroms, J.; Wegscheider, W.; Weiss, D.; Prettl, W. Conversion of Spin into Directed Electric Current in Quantum Wells. *Phys. Rev. Lett.* **2001**, *86*, 4358. [CrossRef] [PubMed]
24. Diehl, H.; Shalygin, V.A.; Bel'kov, V.V.; Hoffmann, C.; Danilov, S.N.; Herrle, T.; Tarasenko, S.A.; Schuh, D.; Gerl, C.; Wegscheider, W.; et al. Spin photocurrents in (110)-grown quantum well structures. *New J. Phys.* **2007**, *9*, 349. [CrossRef]
25. He, X.W.; Shen, B.; Chen, Y.H.; Zhang, Q.; Han, K.; Yin, C.M.; Tang, N.; Xu, F.J.; Tang, C.G.; Yang, Z.J.; et al. Anomalous Photogalvanic Effect of Circularly Polarized Light Incident on the Two-Dimensional Electron Gas in $Al_xGa_{1-x}N$/GaN Heterostructures at Room Temperature. *Phys. Rev. Lett.* **2008**, *101*, 147402. [CrossRef]
26. Yang, C.L.; He, H.T.; Ding, L.; Cui, L.J.; Zeng, Y.P.; Wang, J.N.; Ge, W.K. Spectral Dependence of Spin Photocurrent and Current-Induced Spin Polarization in an InGaAs/InAlAs Two-Dimensional Electron Gas. *Phys. Rev. Lett.* **2006**, *96*, 186605. [CrossRef] [PubMed]
27. Plank, H.; Ganichev, S.D. A review on terahertz photogalvanic spectroscopy of Bi_2Te_3- and Sb_2Te_3-based three dimensional topological insulators. *Solid State Electron.* **2018**, *147*, 44. [CrossRef]
28. McIver, J.W.; Hsieh, D.; Steinberg, H.; Jarillo-Herrero, P.; Gedik, N. Control over topological insulator photocurrents with light polarization. *Nat. Nanotechnol.* **2011**, *7*, 96. [CrossRef]

29. Pan, Y.; Wang, Q.Z.; Yeats, A.L.; Pillsbury, T.; Flanagan, T.C.; Richardella, A.; Zhang, H.J.; Awschalom, D.D.; Liu, C.X.; Samarth, N. Helicity dependent photocurrent in electrically gated $(Bi_{1-x}Sb_x)_2Te_3$ thin films. *Nat. Commun.* **2017**, *8*, 1037. [CrossRef] [PubMed]
30. Yu, J.L.; Zhu, K.J.; Zeng, X.L.; Chen, L.; Chen, Y.H.; Liu, Y.; Yin, C.M.; Cheng, S.Y.; Lai, Y.F.; Huang, J.; et al. Helicity-dependent photocurrent of the top and bottom Dirac surface states of epitaxial thin films of three-dimensional topological insulators Sb_2Te_3. *Phys. Rev. B* **2019**, *100*, 235108. [CrossRef]
31. Yu, J.L.; Zeng, X.L.; Zhang, L.G.; He, K.; Cheng, S.Y.; Lai, Y.F.; Huang, W.; Chen, Y.H.; Yin, C.M.; Xue, Q.K. Photoinduced Inverse Spin Hall Effect of Surface States in the Topological Insulator Bi_2Se_3. *Nano Lett.* **2017**, *17*, 7878. [CrossRef]
32. Gao, Y.; Kaushik, S.; Philip, E.J.; Li, Z.; Qin, Y.; Liu, Y.P.; Zhang, W.L.; Su, Y.L.; Chen, X.; Weng, H.; et al. Chiral terahertz wave emission from the Weyl semimetal TaAs. *Nat. Commun.* **2020**, *11*, 720. [CrossRef]
33. Sun, K.; Sun, S.-S.; Wei, L.-L.; Guo, C.; Tian, H.-F.; Chen, G.-F.; Yang, H.-X.; Li, J.-Q. Circular Photogalvanic Effect in the Weyl Semimetal TaAs. *Chin. Phys. Lett.* **2017**, *34*, 117203. [CrossRef]
34. Sirica, N.; Tobey, R.I.; Zhao, L.X.; Chen, G.F.; Xu, B.; Yang, R.; Shen, B.; Yarotski, D.A.; Bowlan, P.; Trugman, S.A.; et al. Tracking ultrafast photocurrents in the Weyl semimetal TaAs using THz emission spectroscopy. *Phys. Rev. Lett.* **2019**, *122*, 197401. [CrossRef] [PubMed]
35. Leppenen, N.V.; Ivchenko, E.L.; Golub, L.E. Circular Photocurrent in Weyl Semimetals with Mirror Symmetry. *J. Exp. Theor. Phys.* **2019**, *129*, 139. [CrossRef]
36. Golub, L.E.; Ivchenko, E.L. Circular and magnetoinduced photocurrents in Weyl semimetals. *Phys. Rev. B* **2018**, *98*, 075305. [CrossRef]
37. de Juan, F.; Grushin, A.G.; Morimoto, T.; Moore, J.E. Quantized circular photogalvanic effect in Weyl semimetals. *Nat. Commun.* **2017**, *8*, 15995. [CrossRef] [PubMed]
38. Ji, Z.; Liu, G.; Addison, Z.; Liu, W.; Yu, P.; Gao, H.; Liu, Z.; Rappe, A.M.; Kane, C.L.; Mele, E.J.; et al. Spatially dispersive circular photogalvanic effect in a Weyl semimetal. *Nat. Mater.* **2019**, *18*, 955. [CrossRef] [PubMed]
39. Chan, C.K.; Lindner, N.H.; Refael, G.; Lee, P.A. Photocurrents in Weyl semimetals. *Phys. Rev. B* **2017**, *95*, 041104. [CrossRef]
40. Karch, J.; Olbrich, P.; Schmalzbauer, M.; Brinsteiner, C.; Wurstbauer, U.; Glazov, M.M.; Tarasenko, S.A.; Ivchenko, E.L.; Weiss, D.; Eroms, J.; et al. Photon helicity driven electric currents in graphene. *arXiv* **2010**, arXiv:1002.1047.
41. Jiang, C.; Shalygin, V.A.; Panevin, V.Y.; Danilov, S.N.; Glazov, M.M.; Yakimova, R.; Lara-Avila, S.; Kubatkin, S.; Ganichev, S.D. Helicity-dependent photocurrents in graphene layers excited by midinfrared radiation of a CO_2 laser. *Phys. Rev. B* **2011**, *84*, 125429. [CrossRef]
42. Glazov, M.M.; Ganichev, S.D. High frequency electric field induced nonlinear effects in graphene. *Phys. Rep.* **2014**, *535*, 101. [CrossRef]
43. Ganichev, S.D.; Weiss, D.; Eroms, J. Terahertz Electric Field Driven Electric Currents and Ratchet Effects in Graphene. *Ann. Phys.* **2017**, *529*, 1600406. [CrossRef]
44. Qian, X.; Cao, B.; Wang, Z.; Shen, X.; Soci, C.; Eginligil, M.; Yu, T. Carrier density and light helicity dependence of photocurrent in mono- and bilayer graphene. *Semicond. Sci. Technol.* **2018**, *33*, 114008. [CrossRef]
45. Hosur, P. Circular photogalvanic effect on topological insulator surfaces: Berry-curvature-dependent response. *Phys. Rev. B* **2011**, *83*, 035309. [CrossRef]
46. Junck, A.; Refael, G.; Oppen, F.V. Photocurrent response of topological insulator surface states. *Phys. Rev. B* **2013**, *88*, 075144. [CrossRef]
47. Liang, G.; Zhai, G.; Ma, J.; Wang, H.; Zhao, J.; Wu, X.; Zhang, X. Strain-induced circular photogalvanic current in Dirac semimetal Cd_3As_2 film epitaxied on GaAs (111)B substrate. *Nanoscale* **2022**, *14*, 2383–2392. [CrossRef]
48. Wang, B.M.; Zhu, Y.; Travaglini, H.C.; Savrasov, S.Y.; Yu, D. Schottky Electric Field Induced Circular Photogalvanic Effect in Cd_3As_2 Nanobelts. *arXiv* **2022**, arXiv:2210.03819.
49. Young, S.M.; Zaheer, S.; Teo, J.C.; Kane, C.L.; Mele, E.J.; Rappe, A.M. Dirac Semimetal in Three Dimensions. *Phys. Rev. Lett.* **2012**, *108*, 140405. [CrossRef]
50. Pan, H.; Wu, M.; Liu, Y.; Yang, S.A. Electric control of topological phase transitions in Dirac semimetal thin films. *Sci. Rep.* **2015**, *5*, 14639. [CrossRef] [PubMed]
51. Shao, D.; Ruan, J.; Wu, J.; Chen, T.; Guo, Z.; Zhang, H.; Sun, J.; Sheng, L.; Xing, D. Strain-induced quantum topological phase transitions in Na_3Bi. *Phys. Rev. B* **2017**, *96*, 075112. [CrossRef]
52. Jin, H.; Dai, Y.; Ma, Y.D.; Li, X.; Wei, W.; Yu, L.; Huang, B. The electronic and magnetic properties of transition-metal element doped three-dimensional topological Dirac semimetal Cd_3As_2. *J. Mater. Chem. C* **2015**, *3*, 3547. [CrossRef]
53. Liu, Y.; Tiwari, R.; Narayan, A.; Jin, Z.; Yuan, X.; Zhang, C.; Chen, F.; Li, L.; Xia, Z.; Sanvito, S.; et al. Cr doping induced negative transverse magnetoresistance in Cd_3As_2 thin films. *Phys. Rev. B* **2018**, *97*, 085303. [CrossRef]
54. Yuan, X.; Chen, P.H.; Zhang, L.Q.; Zhang, C.; Wang, J.Y.; Liu, Y.W.; Sun, Q.; Zhou, P.; Zhang, D.W.; Hu, Z.G.; et al. Direct observation of landau level resonance and mass generation in Dirac semimetal Cd_3As_2 thin films. *Nano Lett.* **2017**, *17*, 2211. [CrossRef]
55. Zakhvalinskii, V.S.; Nikulicheva, T.B.; Lähderanta, E.; Shakhov, M.A.; Nikitovskaya, E.A.; Taran, S.V. Anomalous cyclotron mass dependence on the magnetic field and Berry's phase in $(Cd_{1-x-y}Zn_xMn_y)_3As_2$ solid solutions. *J. Phys. Condens. Matter* **2017**, *29*, 455701. [CrossRef]

56. Sun, Y.; Meng, Y.; Dai, R.; Yang, Y.; Xu, Y.; Zhu, S.; Shi, Y.; Xiu, F.; Wang, F. Slowing down photocarrier relaxation in Dirac semimetal Cd_3As_2 via Mn doping. *Opt. Lett.* **2019**, *44*, 4103–4106. [CrossRef] [PubMed]
57. Wang, H.; Ma, J.; Wei, Q.; Zhao, J. Mn doping effects on the gate-tunable transport properties of Cd_3As_2 films epitaxied on GaAs. *J. Semicond.* **2020**, *41*, 072903. [CrossRef]
58. Ma, J.L. Molecular-Beam Epitaxy Growth and Physical Properties of Mn-Based Spintronic Materials. Ph.D. Thesis, University of Chinese Academy of Sciences, Beijing, China, 2019.
59. Zhai, G.H. The Carrier Dynamics Study of Dirac Semimetal Cd_3As_2. Ph.D. Thesis, University of Chinese Academy of Sciences, Beijing, China, 2021.
60. Hase, M.; Ishioka, K.; Demsar, J.; Ushida, K.; Kitajima, M. Ultrafast dynamics of coherent optical phonons and nonequilibrium electrons in transition metals. *Phys. Rev. B* **2005**, *71*, 184301. [CrossRef]
61. Yu, J.; Cheng, S.; Lai, Y.; Zheng, Q.; Zhu, L.; Chen, Y.; Ren, J. Temperature dependence of spin photocurrent spectra induced by Rashba- and Dresselhaus-type circular photogalvanic effect at inter-band excitation in InGaAs/AlGaAs quantum wells. *Opt. Express* **2015**, *23*, 27250–27259. [CrossRef]
62. Wang, Y.M.; Yu, J.L.; Zeng, X.L.; Chen, Y.H.; Liu, Y.; Cheng, S.Y.; Lai, Y.F.; Yin, C.M.; He, K.; Xue, Q.K. Temperature and excitation wavelength dependence of circular and linear photogalvanic effect in a three dimensional topological insulator Bi_2Se_3. *J. Phys. Condens. Matter* **2019**, *31*, 415702. [CrossRef]
63. Eldridge, P.S.; Leyland, W.; Lagoudakis, P.G.; Karimov, O.Z.; Henini, M.; Taylor, D.; Phillips, R.T.; Harley, R.T. All-optical measurement of Rashba coefficient in quantum wells. *Phys. Rev. B* **2008**, *77*, 125344. [CrossRef]
64. Zhai, G.; Ma, J.; Wang, H.; Ye, J.; Li, T.; Li, Y.; Liang, G.; Zhao, J.; Wu, X.; Zhang, X. Photoexcited carrier dynamics of thin film Cd_3As_2 grown on a GaAs(111)B substrate by molecular beam epitaxy. *Phys. Rev. B* **2021**, *104*, 094302. [CrossRef]
65. Yang, Y.; Zhou, J.; Xie, X.; Zhang, X.; Li, Z.; Liu, S.; Ai, L.; Ma, Q.; Leng, P.; Zhao, M.; et al. Photodetection and Infrared Imaging Based on Cd_3As_2 Epitaxial Vertical Heterostructures. *ACS Nano* **2022**, *16*, 12244–12252. [CrossRef] [PubMed]
66. Lu, W.; Ge, S.; Liu, X.; Lu, H.; Li, C.; Lai, J.; Zhao, C.; Liao, Z.; Jia, S.; Sun, D. Ultrafast relaxation dynamics of photoexcited Dirac fermions in the three-dimensional Dirac semimetal Cd_3As_2. *Phys. Rev. B* **2017**, *95*, 024303. [CrossRef]

nanomaterials

MDPI

Article

Size-Dependent Superconducting Properties of in Nanowire Arrays

Alexey A. Noyan [1,2], Yevgeniy A. Ovchenkov [2], Valery V. Ryazanov [3,4], Igor A. Golovchanskiy [1,4], Vasily S. Stolyarov [1,4], Eduard E. Levin [2] and Kirill S. Napolskii [2,4,*]

1 Moscow Institute of Physics and Technology, 141700 Dolgoprudny, Russia
2 Lomonosov Moscow State University, 119991 Moscow, Russia
3 Institute of Solid State Physics RAS, 142432 Chernogolovka, Russia
4 National University of Science and Technology MISIS, 119049 Moscow, Russia
* Correspondence: kirill@inorg.chem.msu.ru

Abstract: Arrays of superconducting nanowires may be useful as elements of novel nanoelectronic devices. The superconducting properties of nanowires differ significantly from the properties of bulk structures. For instance, different vortex configurations of the magnetic field have previously been predicted for nanowires with different diameters. In the present study, arrays of parallel superconducting In nanowires with the diameters of 45 nm, 200 nm, and 550 nm—the same order of magnitude as coherence length ξ—were fabricated by templated electrodeposition. Values of magnetic moment M of the samples were measured as a function of magnetic field H and temperature T in axial and transverse fields. $M(H)$ curves for the arrays of nanowires with 45 nm and 200 nm diameters are reversible, whereas magnetization curves for the array of nanowires with 550 nm diameter have several feature points and show a significant difference between increasing and decreasing field branches. Critical fields increase with a decrease in diameter, and the thinnest nanowires exceed bulk critical fields by 20 times. The qualitative change indicates that magnetic field configurations are different in the nanowires with different diameters. Variation of $M(H)$ slope in small fields, heat capacity, and the magnetic field penetration depth with the temperature were measured. Superconductivity in In nanowires is proven to exist above the bulk critical temperature.

Keywords: superconducting nanowires; anodic aluminum oxide; indium; electrodeposition; magnetization curve

check for updates

Citation: Noyan, A.A.; Ovchenkov, Y.A.; Ryazanov, V.V.; Golovchanskiy, I.A.; Stolyarov, V.S.; Levin, E.E.; Napolskii, K.S. Size-Dependent Superconducting Properties of in Nanowire Arrays. *Nanomaterials* **2022**, *12*, 4095. https://doi.org/10.3390/nano12224095

Academic Editors: Gang Xiang and Hongtao Ren

Received: 20 October 2022
Accepted: 14 November 2022
Published: 21 November 2022

1. Introduction

The development of superconducting nanoelectronic devices has remained an important task of studies in recent years. Key elements of the devices are Josephson junctions with tailored properties [1] and elements based on superconducting weak links, such as SQUID [2,3] and quantum interference effect transistors [4]. There is a constant endeavor to improve the properties of superconducting devices: the increase in breakdown current [5] and stability in the air [6], the rise in working temperature [7], and the increase in magnetic field sensibility [8]. Improvement potential is limited for nanostructures fabricated by sputtering, as this technology has significant limitations in controlling the crystallinity of the samples. For instance, magnetron sputtering commonly produces fine crystalline films with a grain size of less than several nanometers. Morphology of the sputtered films may change significantly later, for example, in the case of thermal treatment, but additional difficulties with the stability of chemical composition and morphology of all elements in the device at high temperatures arise. Therefore, the use of coarse-crystalline superconducting nanowires obtained by templated electrodeposition [7–12] as the weak link is a possible step to improve superconducting devices. The templated electrodeposition has been successfully used for many metals and is being actively developed [13–18]. If deposition parameters are adjusted properly, the fabrication of single-crystalline nanowires is possible. Thereby Josephson junctions based on electrodeposited nanowires have superior characteristics compared to junctions fabricated by sputtering [7]. Other advantages of electrodeposition are a wide range of deposition solutions and the ability to obtain multi-segmented nanowires [19].

Studies of superconductivity In nanowires are essential for assembling nanoelectronic devices. Nanowires have shown several unique features of electron transport; one of the questions is the interplay of superconductivity and ferromagnetism In nanowires. The works [12,20] showed a decrease in resistance in 1-µm-long Co nanowires used as a weak link. That is a great length compared to a scale of 1 nm magnitude order at which proximity effect is observed in bulk ferromagnets. In [21], it was shown that Pb nanowire with the admixture of Co exhibits both superconducting and ferromagnetic properties. Another peculiar feature is the superconductivity of single-crystalline Bi nanowires at temperatures up to 1.3 K [16], whereas the critical temperature of the bulk Bi is as low as 0.53 mK [22].

Magnetic field distribution in superconducting nanowires is another open question. Three states of superconductor are distinguished: Meissner state, where the magnetic field is completely expelled, vortex state with a lattice of single quantum vortices, and intermediate mixed state in which more sophisticated field configurations are possible [23]. The picture is complicated in the case of nanowires because of geometrical limitations. In [24], a numerical simulation was conducted for nanowires of a type I superconductor in the transverse magnetic field. Diameters of the nanowires from 5 ξ_{0_b} to 50 ξ_{0_b} were studied, where ξ_{0_b} is bulk coherence length. The simulation showed that nanowires could demonstrate type-II magnetic response. Different distributions of the penetrated field In nanowires depending on the diameter are predicted; the configurations differ significantly from ones observed in bulk superconductors and thin films [25]. Modeled dependence of magnetization M as a function of magnetic field strength H is nonmonotonic with several features related to shifts between different configurations. Two vortex configurations of the penetrated magnetic field were observed in simulation for the thinnest nanowires analyzed (Figure 1): 1D lattice of single-quantum vortices (a) and a row of single-quantum vortices divided into pairs (b). It is worth noting that the theoretical results [24] have not been proved experimentally yet.

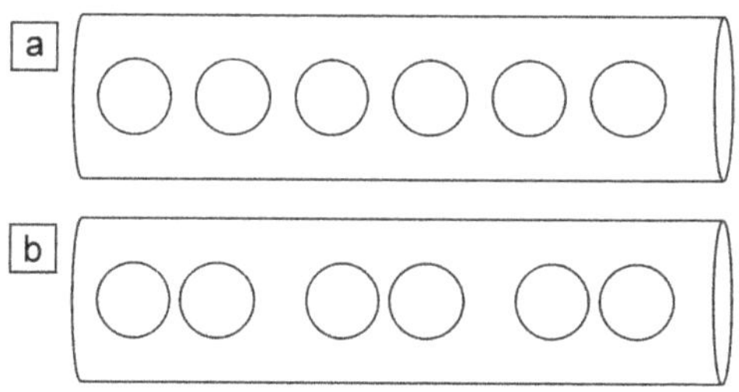

Figure 1. Scheme of the simplest possible vortex configurations In nanowire according to the simulation [24]: 1D lattice of single-quantum vortices (**a**) and a row of single-quantum vortices divided into pairs (**b**). Circles are superconducting vortices.

Magnetization behavior of β-Ga nanowire array with a diameter of 140 nm (2.3 ξ_{0_b}) in the transverse field was studied experimentally [26]. $M(H)$ curves demonstrated hysteresis and type-II-like behavior. The authors interpreted the observed $M(H)$ dependences as the sign of vortex configuration illustrated in Figure 1a. Magnetization of a single Pb nanowire with 390 nm (4.5ξ_{0_b}) diameter was measured locally by the Hall probe [27]. $M(H)$ curve had hysteresis at low temperatures, which disappeared at temperatures closer to the T_C. The simulation was made for low temperatures, which showed the configuration of the penetrated field as a lattice of single-quantum vortices similar to the configuration in Figure 1a. Simulated data agreed with the experimental results. Magnetization $M(H)$ for

Pb nanowires with 200 nm (2.3 ξ_{0_b}) diameter were measured in [28]. The studies [27,28] showed the enhancement of critical magnetic fields In nanowires.

In the present work, arrays of In nanowires were analyzed. Specific properties of In are large coherence length and the presence of a thin (about 4 nm) surface oxide layer preventing the metal core from further oxidation in the air [29]. These properties make In promising for applications in superconducting electronics. Bulk In is a type I superconductor with a critical temperature T_C = 3.408 K and a zero-temperature critical field $H_C(0)$ = 281.5 Oe [30]. The mean value of coherence length measured in bulk indium is ξ_{0_b} = 264 nm, and the mean penetration depth is λ_{0_b} = 53 nm [31–37]; all values of λ_{0_b} and ξ_{0_b}, which were averaged, are listed in Appendix A. Experimentally observed temperature dependence of critical field H_C is in good agreement with the general law [30]:

$$H_C(T) = H_C(0)\left(1 - \left(\frac{T}{T_C}\right)^2\right) \tag{1}$$

Different methods to fabricate In nanowires were proposed in studies [38–49], including vacuum sputtering in V-shaped silica grooves [40,41], oblique angle deposition [42], and injection of molten metal into a porous template under pressure [48]. Templated electrodeposition of In nanowires [38,39,49] is the most promising method as it allows one to set geometrical parameters of the nanowires precisely and, at the same time, to fabricate single-crystalline nanowires after adjustment of deposition conditions. To the best of our knowledge, superconductivity in In nanowires has not been studied yet. However, several studies on In nanoparticles have been conducted [50,51], where molten indium was injected into the porous glass under pressure. The characteristic size of the particles was determined as a function of the pressure used, but the shape of the particles and size distribution were not defined. The measurements of magnetization curves showed a significant increase in critical fields with a decrease in indium particle size. Additionally, the rise of critical temperature was observed compared to the bulk In for the particles with a characteristic size of 3.1 nm: the critical temperature was 0.75 K higher than the bulk critical temperature. For particles with a size of 25 nm, the increase in critical temperature within experimental error was measured.

2. Materials and Methods

Arrays of superconducting In nanowires were fabricated by electrodeposition technique using polycarbonate track-etched membranes (Whatman, Little Chalfont, UK; pore diameter of 550 nm, thickness of 20 microns) and anodic aluminium oxide (AAO) as templates. Porous AAO films were prepared by anodization of a high-purity aluminium foil (99.99%) with a thickness of 100 μm. Prior to anodization, the foil was electrochemically polished to a mirror finish in a solution containing 13 M H_3PO_4 and 1.85 M CrO_3 at 80 °C as described elsewhere [52]. The aluminium electrode was polarized 40 times for 3 s at an anodic current density of 0.5 A cm^{-2} with an interpulse interval of 40 s. Two series of AAO templates with pore diameters of 45 nm and 200 nm were fabricated in different anodizing regimes. The film with a pore diameter of 45 nm was formed using two-step anodization procedure in 0.3 M oxalic acid (Chimmed, Moscow, Russia) at a voltage of 40 V and an electrolyte temperature of 0 °C. During the first anodization step, a 10 μm thick sacrificial alumina layer was formed. This layer was then selectively dissolved in an aqueous solution containing 0.5 M H_3PO_4 and 0.2 M CrO_3 at 70 °C for 30 min. The second anodization under the same conditions as the first one was stopped when AAO thickness reached 40 μm. The AAO films with a pore diameter of 200 nm and a thickness of 46 μm were formed in 0.1 M phosphoric acid (Chimmed, Moscow, Russia) at a voltage of 195 V and an electrolyte temperature of 0 °C. After anodizing, the porous oxide films were washed repeatedly in deionized water and dried in air. The residual Al was dissolved in a solution of Br_2 in CH_3OH (1:10 volume). The barrier oxide layer was removed by chemical etching in 3 M H_3PO_4 solution at room temperature with the electrochemical detection of the pore opening moment [53].

Electrodeposition of In was performed in a three-electrode cell in potentiostatic mode using a PGSTAT100N (Metrohm Autolab, Utrecht, The Netherlands) instrument. AAO and track-etched membranes were coated with a 300-nm-thick Au layer using a magnetron sputtering system. This continuous gold layer served as a current collector. Before In electrodeposition 2 μm long Au nanorods were deposited to prevent the formation of superconducting material under a porous template. For this purpose, a commercial electrolyte (Ecomet, Moscow, Russia) containing 0.04 M $Au[CN]_2^-$ in citrate buffer (pH = 6) was used. A Pt wire ring with a diameter of ca. 2 cm was used as a counter electrode. The saturated (KCl) Ag/(AgCl) electrode connected with the cell via Luggin capillary served as the reference electrode; deposition potential was −1.0 V. Indium was deposited from a solution containing 0.1 M $In(SO_3NH_2)_3$ and 1.25 M $Na(SO_3NH_2)$ (pH = 2.1). An indium rod served as the reference electrode. An indium ring was used as a counter electrode. A deposition potential of −0.4 V was used. To increase the filling factor during the electrodeposition of In in AAO films the template surface was cleaned repeatedly by electrolyte flow. Indium in track-etched membranes was deposited to half of the membrane thickness. After the electrodeposition, the top side of the AAO films and the current collector side of the track-etched membrane were mechanically polished to remove indium from the surface (Figure 2). This procedure allows one to obtain samples with metal solely in the pores.

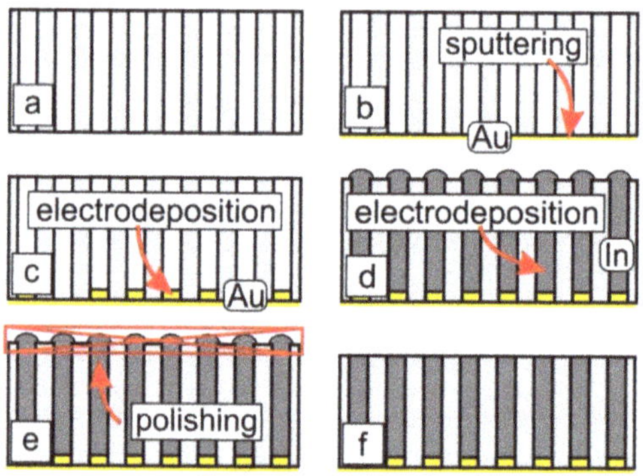

Figure 2. A flowchart of In nanowire array preparation process. (**a**) The original porous template. (**b**) Sputtering of Au layer on the porous template. (**c**) Templated electrodeposition of short Au nanorods. (**d**) Templated electrodeposition of In until complete filling of the pores by the metal. (**e**) Mechanical polishing of the top part of the nanocomposite to remove In from sample surface. (**f**) Resulting nanocomposite.

Magnetization measurements were carried out using an MPMS XL (Quantum Design, Darmstadt, Germany) SQUID magnetometer. The phase composition of the samples was characterized with a Aeris (Malvern Panalytical, Malvern, UK) tabletop X-ray diffractometer (CuKα radiation, PIXcel3D detector). The morphology of the porous oxide films and track-etched membranes was characterized using a NVision 40 (Carl Zeiss, Oberkochen, Germany) scanning electron microscope. Before SEM analysis, the samples were coated with a 5-nm-thick conductive layer of chromium using a Q150T ES (Quorum Technologies, Lewes, UK) sputter coater. The mean porosity of AAO films with a pore diameter of 45 nm was measured using the birefringence method [54].

3. Results and Discussion

Three kinds of superconducting In nanowire arrays with various diameters of the nanostructures were prepared. For this purpose, two types of the AAO films and a polymer track-etched membrane with cylindrical pores aligned perpendicular to the surface were used as templates. The morphology of the templates was characterized using SEM (Figure 3). The parameters of the templates calculated from the statistical analysis of the SEM images are listed in Table 1. AAO films possess 2D hexagonal ordered pore arrangement in the plane of the film and narrow interpore distance distribution. Conversely, pores in the polymer track-etched membrane had no positional order; the positions of the pores are random with few pore overlaps; moreover, deviation of pores from the normal to the surface of the track-etched membrane is possible. The porosity of the sample s45 was measured with the optical birefringence method, which is more precise than SEM [54]. Diameters of In nanowires equal to the pore diameter D_p and had the same order of magnitude as ξ_{0_b} and λ_{0_b}, therefore geometrically limited superconducting states could be studied. Values of pore volume V in samples were calculated from the geometrical parameters (Table 1), taking into account a diameter of the porous electrode of 1.2 cm, identical for s45, s200, and s550. X-ray diffraction showed (Figure 4) that nanowire arrays are featuring preferential orientation along the <100> axis.

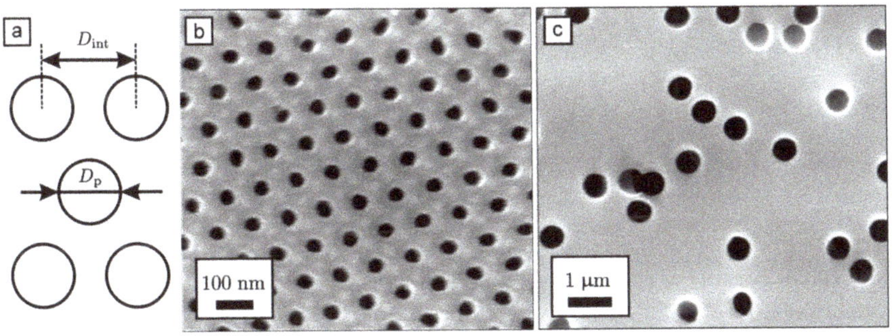

Figure 3. Scheme of pore arrangement (**a**) and SEM images of the surface of AAO film s45 (**b**) and polymer track-etched membrane s550 (**c**).

Table 1. Parameters of AAO films and polymer track-etched membranes, used as templates.

Sample Name	Type of the Template	Interpore Distance D_{int}, nm	Pore Diameter D_p, nm	Pore Density, μm^{-2}	Porosity p, %	Template Thickness, μm	Volume of Pores, 10^{-4} cm^3
s45	AAO	101 ± 6	45	113	18	40	8.1
s200	AAO	534	200	4.05	12.7	46	6.6
s550	track-etched membrane	—	550	0.34	8	20	1.8

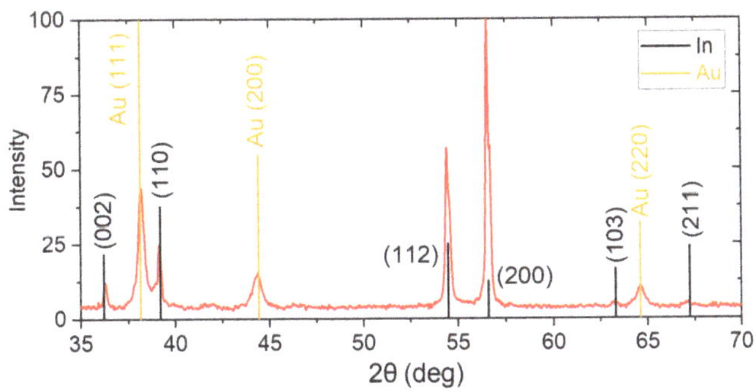

Figure 4. XRD pattern for AAO/In nanocomposite containing array of nanowires with 200 nm diameter.

Figure 5 displays magnetic moment M of In nanowire arrays s45 and s200 as a function of magnetic field H at temperatures from 2 K to 3.3 K. $M(H)$ curves measured for the field perpendicular to the long axis of the nanowires are shown in Figure 5a,b. It is worth noting that all curves measured for the nanowires with diameters of 45 nm and 200 nm are reversible, values of magnetic moment M measured in increasing and decreasing magnetic fields are the same at each field value. The diameter of 200 nm is almost equal to ξ_{0_b} and 45 nm diameter is several times smaller than the coherence length of bulk In. The absence of vortex states in thin nanowires is the probable explanation of reversibility.

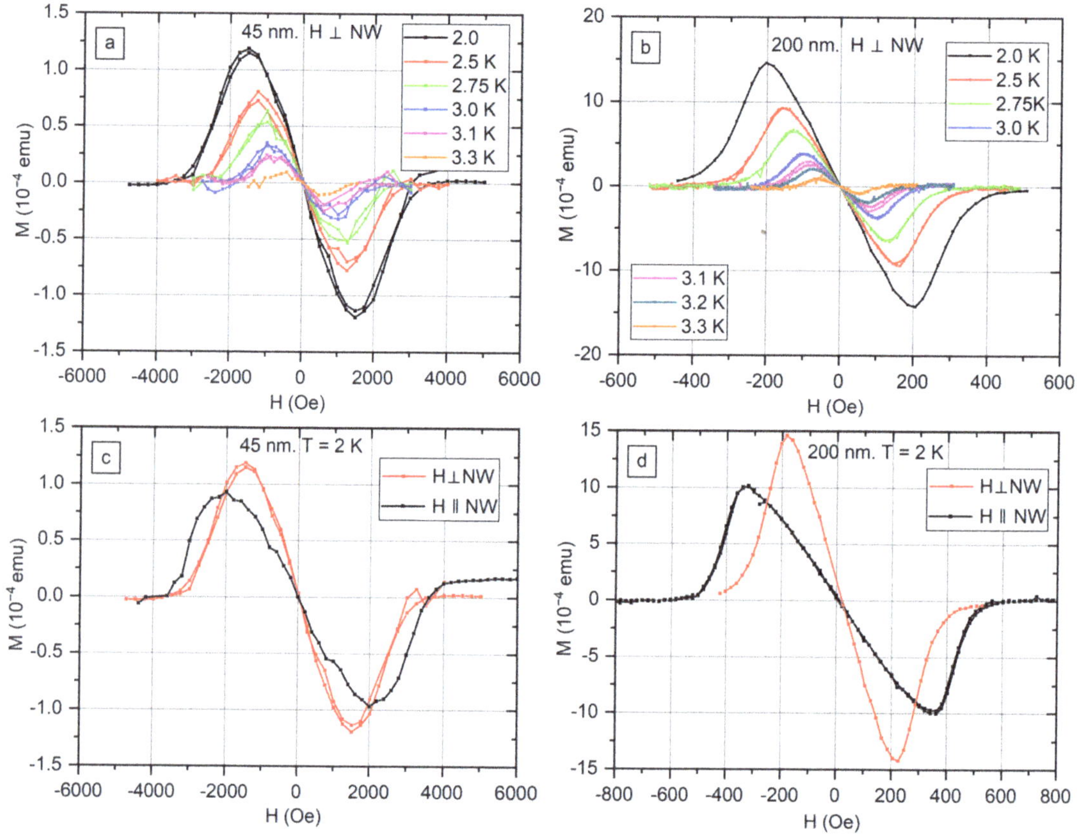

Figure 5. The magnetic moment M as a function of transverse magnetic field H for In nanowire arrays with nanowire diameters 45 nm (**a**) and 200 nm (**b**). The comparison of $M(H)$ curves measured at 2 K temperature for the same sample in transvers and axial fields: s45 (**c**) and s200 (**d**).

Figure 5c,d compares curves measured in transverse and axial fields at 2 K. Curves measured in transverse fields have larger maximum values of M and less sharp decline after the maximum. A different demagnetization factor is one of the reasons for such a difference in the shape of magnetization curves. The demagnetization factor of the cylinder in the axial field is 0, and the demagnetization factor in the transverse field is $\frac{1}{2}$. The macroscopic shape of the sample is another factor that influences the shape of the magnetization curve.

$M(H)$ curves measured for nanowires parallel to the field H at temperatures from 2.0 K to 3.6 K are shown in Figures S1 and S2. Weak diamagnetic response is visible on the $M(H)$ curve for s200 at T = 3.5 K with no response at T = 3.6 K. Therefore, superconductivity in In nanowires is possible at temperatures above a bulk critical temperature of In (T_C = 3.408 K).

The magnetization curves for arrays of nanowires with 550 nm diameter recorded in transverse and axial fields are shown in Figure 6a and 6b, respectively. For a clear view, only the decreasing field branch is presented in Figure 6b. In contrast to s45 and s200 samples, $M(H)$ curves for s550 show hysteresis. Higher magnetic moments are reached in the increasing field branch, and M remains zero in a larger field interval on the decreasing field branch. However, no hysteresis is observed in small fields. The exit of a magnetic field from the nanowire is suppressed. The probable explanation of the observed behavior is the formation of a vortex state In nanowires with a surface barrier causing the hysteresis. Another explanation may be the pinning of vortices on grain boundaries. It should be noted that $M(H)$ curves for s550 in the transverse field are qualitatively similar to measured ones for the array of β-Ga in [26].

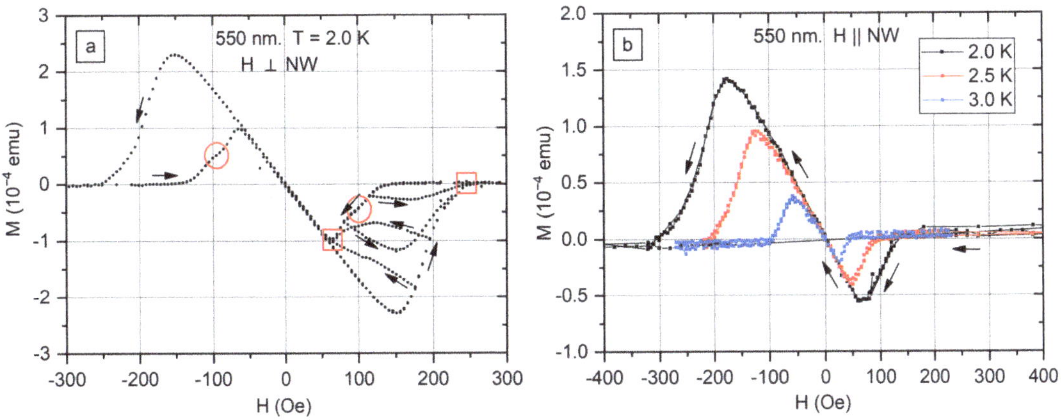

Figure 6. The magnetic moment M of 550-nm-diameter In nanowire array as a function of transverse (**a**) and axial (**b**) magnetic field H. Arrows show how the field was changing. Two joining points of minor loops are marked with red squares. The feature of the $M(H)$ curve in the transverse field, which is absent in case of axial field, is marked with the red circle in panel (**a**).

Figure 6a shows a set of minor loops, directions of field H changing are shown with arrows. All the minor loops for increasing fields join with the main loop at the same point, in which $M = 0$. All the minor loops for decreasing fields join the main loop in another point, corresponding to the maximum magnetization on decreasing field branch. Two joining points of minor loops are marked with squares (Figure 6a).

It is worth noting that the feature present in both positive and negative magnetic fields on the $M(H)$ curve for s550 in the transverse field, where the derivative changes twice (see red circle in Figure 6a). It was observed on curves recorded at temperatures 2.0 K and 2.5 K (Figure S3). The feature corresponds to approximately $\frac{1}{2}$ of the maximum M value. No such features are present on the $M(H)$ curves in the axial field (Figure 6b). A possible explanation of the observed feature is a transition between the two simplest vortex configurations (Figure 1), predicted in the simulations in [24].

There are three critical fields H_m, H_a, and H_d on the $M(H)$ curves that were measured as shown in Figure 7a. H_m is the field of the maximum of magnetization, H_a is the extrapolation of the linear region of M decline, and H_d is the field at which M becomes zero. Figure 7b–d presents characteristic fields for s45, s200, and s550 samples. Values of the critical fields for the array of In nanowires with a diameter of 550 nm were calculated separately for increasing (+) and decreasing (−) branches.

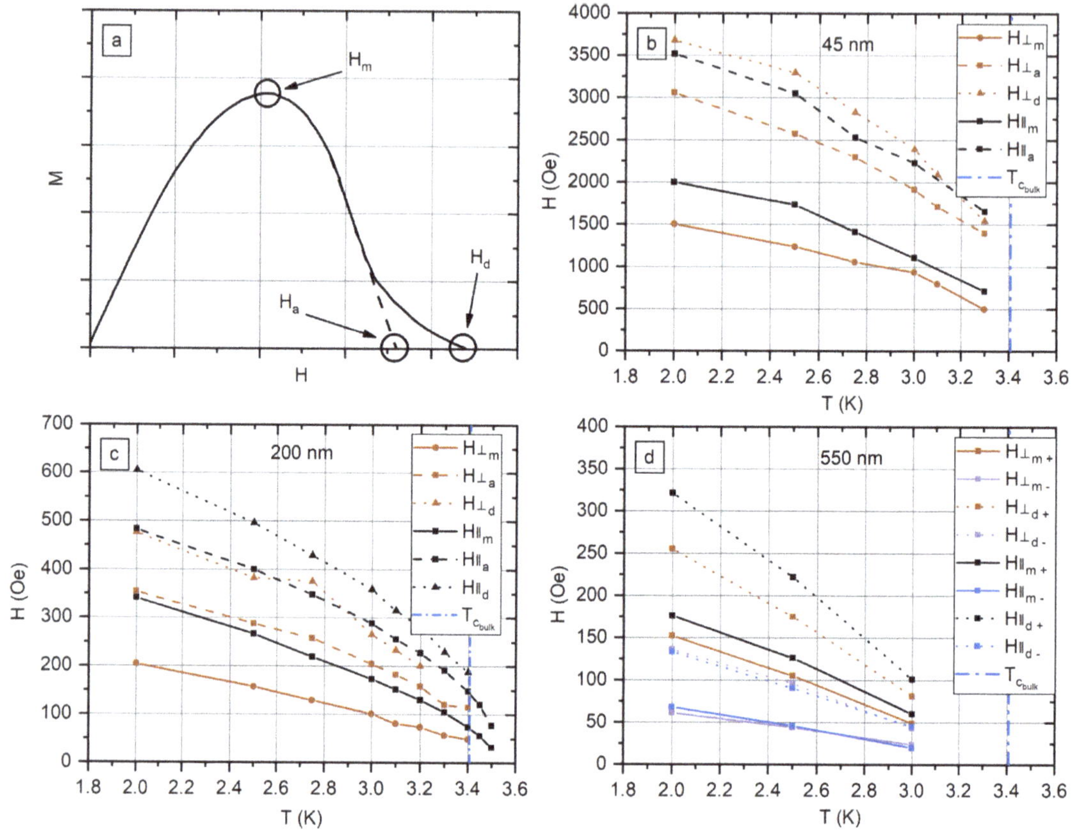

Figure 7. (**a**) Schematic illustration of the critical fields H_m, H_a, and H_d on $M(H)$ curve. Values of fields measured for the arrays of In nanowires with a diameter of 45 nm (**b**), 200 nm (**c**), and 550 nm (**d**). Values measured for increasing in modulus field H have index +, whereas values measured for decreasing in modulus field H have index −.

The analysis of the critical fields shows that superconductivity in In nanowires is possible in much higher fields than in bulk indium. H_m in 550 nm nanowire array are close to the bulk critical field of In, H_{C_b} (2 K) = 180 Oe (see Appendix A), H_d exceeds bulk critical field. The critical fields increase with the decrease In nanowire diameter. In the transverse field, H_m is 200 Oe in 200 nm nanowires and 1500 Oe in 45 nm nanowires. The superconducting response is detected in the field of more than 3500 Oe, exceeding the bulk critical field by 20 times. The shape of the $M(H)$ curve with a slow decay in high fields indicates residual surface superconductivity.

$M(H)$ curves measured for s550 at temperatures 2 K, 2.5 K, and 3.0 K (Figure 5) coincide on the linear region in small fields. The nanowires with a 550 nm diameter are thick enough to perform the Meissner effect with full expulsion of the magnetic field, and the slope is adjusted by the volume of the superconductor. In contrast to the s550 sample, the slope of the linear $M(H)$ region in the s45 and s200 samples depends on the temperature. Figure 8 shows the slopes calculated for s45 and s200. The metal fraction in the nanocomposites s45 and s200 is close (Table 1); however, the slope differs by two orders of magnitude (Figure 8). The reason is that λ is of the same order of magnitude as the nanowire diameter, and full expulsion of the magnetic field does not take place. Therefore, in s45, the sample magnetic field penetrates almost the full nanowire diameter. The small field region is linear, as the configuration of the magnetic field does not change with rise of the field. λ increases with the rise of temperature, accordingly the slope decreases with temperature.

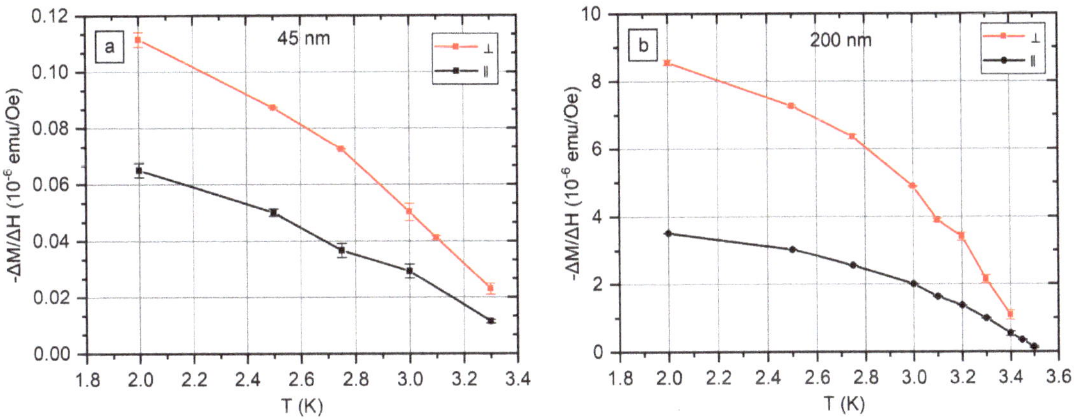

Figure 8. The slope of the linear part of $M(H)$ curve observed at small fields for s45 (**a**) and s200 (**b**).

For nanowires with a diameter smaller than λ in the case of an axial field, the magnetic moment is derived with the equation:

$$\frac{M}{V} = \frac{R^2 H}{32\pi\lambda^2} \tag{2}$$

Magnetic slopes at different temperatures are connected with the equation:

$$\left(\frac{\Delta M}{\Delta H}\right)_1 / \left(\frac{\Delta M}{\Delta H}\right)_2 = \left(\frac{\lambda_2}{\lambda_1}\right)^2, \tag{3}$$

where λ_1 and λ_2 are magnetic field penetration depths at temperatures T_1 and T_2, respectively.

Figure 9a shows λ as a function of T calculated according to (3). Experimental data were fitted with the function:

$$\lambda(T) = \lambda_0 \left(1 - \left(\frac{T}{T_C}\right)^4\right)^{-0.5} \tag{4}$$

The shape of function (4) is in good agreement with the experimental data. Values of T_C are 3.53 K for 200 nm nanowires and 3.46 K for 45 nm nanowires.

Comparing $M(H)$ curves for s45 and s200, the maximum M value for s45 is by one order of magnitude less than the maximum M value for s200. However, the magnetic field related to the maximum for the sample s45 is by one order of magnitude more than the magnetic field for s200.

Therefore, the areas under curves s45 and s200 are of the same order of magnitude. Integral of the moment M over the field is related to the free energy [55]:

$$(F_n - F_s)|_{H=0} = -\int_0^{H_d} M(H)dH, \tag{5}$$

where F_n is the free energy of the sample in normal state, F_s is the free energy in superconducting state.

Using the thermodynamic relations:

$$S = -\frac{\partial F}{\partial T} \tag{6}$$

$$C = T\frac{\partial S}{\partial T} \tag{7}$$

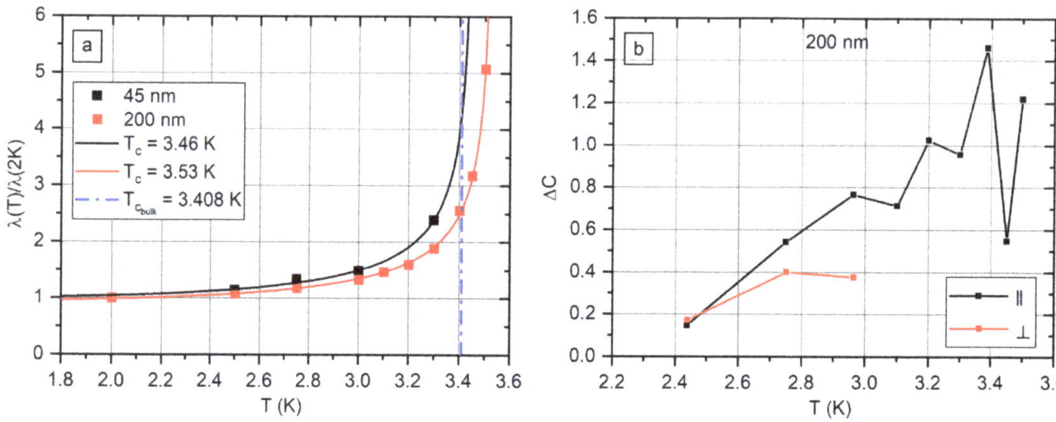

Figure 9. Magnetic field penetration depth as a function of temperature (**a**). The difference in heat capacities ΔC of the normal and the superconducting states as a function of temperature (**b**).

The equation for the heat capacity difference in the normal and superconducting states is as follows:

$$\Delta C = (C_n - C_s)|_{H=0} = T\frac{\partial^2}{\partial T^2}\int_0^{H_d} M(H)dH = T\frac{\partial^2}{\partial T^2}A^M, \tag{8}$$

where A^M is the designation for the integral of M over H:

$$A_i^M = -\int_0^{H_d} M(H,\ T_i)dH \tag{9}$$

If A^M is measured at three temperatures, T_1, T_2, and T_3, the second order derivative may be calculated according to the equations (see scheme of derivation in Figure S4):

$$T = \frac{T_1 + 2T_2 + T_3}{4} \tag{10}$$

$$\frac{\partial^2 A}{\partial T^2}(T) = 2\frac{\frac{A_3^M - A_2^M}{T_3 - T_2} - \frac{A_2^M - A_1^M}{T_2 - T_1}}{T_3 - T_1} \tag{11}$$

Values of A^M were calculated numerically from the $M(H)$ curves. The resulting values of ΔC for the sample s200 are shown in Figure 9b. The extrapolation of data to ΔC = 0 gives T = 2.3 K.

4. Conclusions

Superconducting properties of the arrays of parallel In nanowires with the diameters of 45 nm, 200 nm, and 550 nm were measured using SQUID magnetometry. The diameter of 45 nm is several times smaller than the coherence length ξ_{0_b}, 200 nm is roughly equal to ξ_{0_b}, and 550 nm is several times greater than ξ_{0_b}. The method of arrays fabrication eliminates the possibility of bulk indium existence in the samples; XRD data showed that used deposition conditions result in the formation of crystallographically textured nanowires with preferred orientation along <100> axis.

The qualitative difference was observed in $M(H)$ curves for the nanowires with different diameters. For the arrays of In nanowires with 45 nm and 200 nm diameters, the magnetization curves are completely reversible, increasing field branch is the same as decreasing field branch. For s550 sample, $M(H)$ curves are irreversible. However, no hysteresis is present in a small field. Presumably, the hysteresis is related to vortex states in the nanowires thick enough. Moreover, characteristic points are observed which may be related to shifts between vortex configurations. The emergence of different configurations is possible if more complicated interaction than repulsion exists between vortices.

Therefore, In, which is a type-I superconductor in bulk state, shows type-II-like behavior if formed in 550 nm nanowires. Nanowires with a diameter less than ξ_{0_b} show quasi-type-I behavior: no vortex states are possible because of geometrical limitations. However, the properties of nanowires differ from bulk properties significantly. Critical fields rise with a decrease In nanowire diameter, rise of critical fields up to 20 times compared to the bulk critical field was observed. The increase in critical temperature compared to the bulk values is observed experimentally. The results show wide possibilities for tuning superconducting properties of nanowires by varying diameters and may be useful for applications in nanoelectronics.

Further research in this area could be related to studying the properties of superconducting nanowires in a wider range of diameters to compare the experiment with the theoretical model quantitatively. Another research direction can be aimed to measuring the transport properties of superconducting In nanowires. A change in critical fields by dozens of times should lead to a change in transport properties. Measuring of the transport properties of the structures composed of several In nanowires (e.g., SQUIDs) is also of great interest.

Supplementary Materials: The following supporting information can be downloaded at: https://www.mdpi.com/article/10.3390/nano12224095/s1, Figure S1: The magnetic moment M as a function of axial magnetic field H for the arrays of In nanowires with diameter of 200 nm; Figure S2: The magnetic moment M as a function of axial magnetic field H for the arrays of In nanowires with diameter of 45 nm; Figure S3: The magnetic moment M of the arrays of In nanowires with diameter of 550 nm as a function of magnetic field H applied perpendicular to the nanowires; Figure S4: Scheme of Equation (11) derivation.

Author Contributions: Methodology, investigation, formal analysis, writing—original draft preparation, visualization, data curation, A.A.N.; investigation, Y.A.O.; validation, V.V.R.; investigation, I.A.G.; validation, V.S.S.; investigation, E.E.L.; supervision, writing—review and editing, visualization, project administration, validation, investigation, K.S.N. All authors have read and agreed to the published version of the manuscript.

Funding: The work was performed under financial support of the Russian Science Foundation (grant number 22-23-00984). A.A.N. acknowledges the personal support from the Russian Foundation for Basic Research (grant number 20-33-90084). The theoretical description of the obtained results was performed in frame of the Federal Academic Leadership Program "Priority-2030" (NUST MISIS grant No. K2-2022-029).

Data Availability Statement: The data presented in this study are available on request from the corresponding author.

Acknowledgments: The authors thank Vasiliev A.V. (Department of Chemistry, Lomonosov Moscow State University) for the help with SQUID measurements, Tsirlina G.A. (Department of Chemistry, Lomonosov Moscow State University), and Shanenko A.A. (HSE Tikhonov Moscow Institute of Electronics and Mathematics (MIEM HSE)) for fruitful discussions. SEM images were obtained using the equipment of the IGIC RAS Joint Research Centre for Physical Methods of Research.

Conflicts of Interest: The authors declare no conflict of interest.

Appendix A

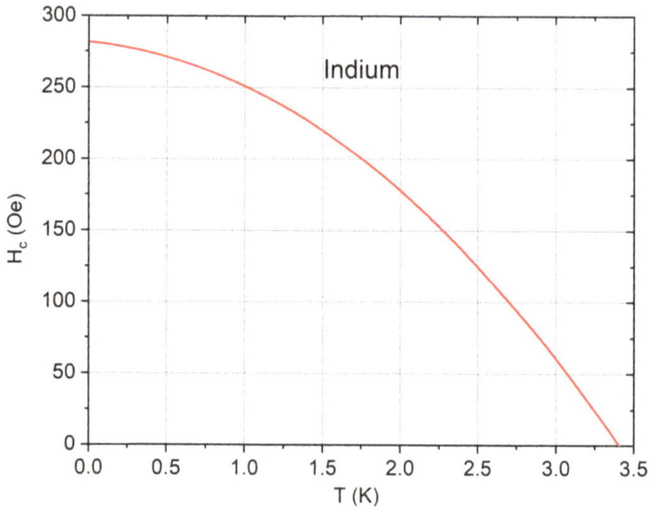

Figure A1. Experimental measurement of the critical field in bulk In as a function of temperature [30].

Table A1. Values of the coherence length and the penetration depth measured in bulk In.

Coherence Length, nm	Penetration Depth, nm	Reference
260		[31]
324		[32]
270		[33]
	70	[34]
	43 ± 2	[35]
200		[36]
	25	[37]

References

1. Morosh, V.; Linek, J.; Müller, B.; Martínez-Pérez, M.J.; Wolter, S.; Weimann, T.; Beyer, J.; Schurig, T.; Kieler, O.; Zorin, A.B.; et al. Transport and Noise Properties of sub-100-nm Planar Nb Josephson Junctions with Metallic Hf-Ti Barriers for nano-SQUID Applications. *Phys. Rev. Appl.* **2020**, *14*, 054072. [CrossRef]
2. Jabdaraghi, R.N.; Peltonen, J.T.; Saira, O.P.; Pekola, J.P. Low-temperature characterization of Nb-Cu-Nb weak links with Ar ion-cleaned interfaces. *Appl. Phys. Lett.* **2016**, *108*, 042604. [CrossRef]
3. Collienne, S.; Raes, B.; Keijers, W.; Linek, J.; Koelle, D.; Kleiner, R.; Kramer, B.G.; Van de Vondel, J.; Silhanek, A.V. Nb-Based Nanoscale Superconducting Quantum Interference Devices Tuned by Electroannealing. *Phys. Rev. Appl.* **2020**, *15*, 034016. [CrossRef]
4. Ronzania, A.; Altimiras, C.; Giazotto, F. Balanced double-loop mesoscopic interferometer based on Josephson proximity nanojunctions. *Appl. Phys. Lett.* **2014**, *104*, 032601. [CrossRef]
5. Kang, C.G.; Lim, S.K.; Lee, S.; Lee, S.K.; Cho, C.; Lee, Y.G.; Hwang, H.J.; Kim, Y.; Choi, H.J.; Choe, S.H. Effects of multi-layer graphene capping on Cu interconnects. *Nanotechnology* **2014**, *24*, 115707. [CrossRef]
6. Xu, W.H.; Wang, L.; Guo, Z.; Chen, X.; Liu, J.; Huang, X.J. Copper Nanowires as Nanoscale Interconnects: Their Stability, Electrical Transport, and Mechanical Properties. *ACS Nano* **2015**, *9*, 241–250. [CrossRef]
7. Skryabina, O.V.; Egorov, S.V.; Goncharova, A.S.; Klimenko, A.A.; Kozlov, S.N.; Ryazanov, V.V.; Bakurskiy, S.V.; Kupriyanov, M.Y.; Golubov, A.A.; Napolskii, K.S.; et al. Josephson coupling across a long single-crystalline Cu nanowire. *Appl. Phys. Lett.* **2017**, *110*, 222605. [CrossRef]
8. De Cecco, A.; Calvez, K.L.; Sacépé, B.; Winkelmann, C.B.; Courtois, H. Interplay between electron overheating and ac Josephson effect. *Phys. Rev. B* **2016**, *93*, 180505. [CrossRef]

9. Kompaniiets, M.; Dobrovolskiy, O.V.; Neetzel, C.; Begun, E.; Porrati, F.; Ensinger, W.; Huth, M. Proximity-induced superconductivity in crystalline Cu and Co nanowires and nanogranular Co structures. *J. Appl. Phys.* **2014**, *116*, 073906. [CrossRef]
10. Jung, M.; Noh, H.; Doh, Y.J.; Song, W.; Chong, Y.; Choi, M.S.; Yoo, Y.; Seo, K.; Kim, N.; Woo, B.C.; et al. Superconducting Junction of a Single-Crystalline Au Nanowire for an Ideal Josephson Device. *ACS Nano* **2011**, *5*, 2271–2276. [CrossRef]
11. Skryabina, O.V.; Kozlov, S.N.; Egorov, S.V.; Klimenko, A.A.; Ryazanov, V.V.; Bakurskiy, S.V.; Kupriyanov, M.Y.; Klenov, N.V.; Soloviev, I.I.; Golubov, A.A.; et al. Anomalous magneto-resistance of Ni-nanowire/Nb hybrid system. *Sci. Rep.* **2019**, *9*, 14470. [CrossRef] [PubMed]
12. Wang, J.; Singh, M.; Tian, M.; Kumar, N.; Liu, B.; Shi, C.; Jain, J.K.; Samarth, N.; Mallouk, T.E.; Chan, M.H.W. Interplay between superconductivity and ferromagnetism in crystalline nanowires. *Nat. Phys.* **2010**, *6*, 389–394. [CrossRef]
13. Thurn-Albrecht, T.; Schotter, J.; Kästle, G.A.; Emley, N.; Shibauchi, T.; Krusin-Elbaum, L.; Guarini, K.; Black, C.T.; Tuominen, M.T.; Russell, T.P. Ultrahigh-density nanowire arrays grown in self-assembled diblock copolymer templates. *Science* **2000**, *290*, 2126–2129. [CrossRef]
14. Chen, Z.; Ye, S.; Stewart, I.E.; Wiley, B.J. Copper Nanowire Networks with Transparent Oxide Shells That Prevent Oxidation without Reducing Transmittance. *ACS Nano* **2014**, *8*, 9673–9679. [CrossRef] [PubMed]
15. Shin, S.; Al-Housseiny, T.T.; Kim, B.S.; Cho, H.H.; Stone, H.A. The Race of Nanowires: Morphological Instabilities and a Control Strategy. *Nano Lett.* **2014**, *14*, 4395–4399. [CrossRef]
16. Tian, M.; Wang, J.; Ning, W.; Mallouk, T.E.; Chan, M.H.W. Surface Superconductivity in Thin Cylindrical Bi Nanowire. *Nano Lett.* **2015**, *15*, 1487–1492. [CrossRef]
17. Leontiev, A.P.; Brylev, O.A.; Napolskii, K.S. Arrays of rhodium nanowires based on anodic alumina: Preparation and electrocatalytic activity for nitrate reduction. *Electrochim. Acta* **2015**, *155*, 466–473. [CrossRef]
18. Napolskii, K.S.; Roslyakov, I.V.; Eliseev, A.A.; Petukhov, D.I.; Lukashin, A.V.; Chen, S.F.; Liu, C.P.; Tsirlina, G.A. Tuning the microstructure and functional properties of metal nanowire arrays via deposition potential. *Electrochim. Acta* **2011**, *56*, 2378–2384. [CrossRef]
19. Malysheva, I.V.; Kolmychek, I.A.; Romashkina, A.M.; Leontiev, A.P.; Napolskii, K.S.; Murzina, T.V. Magneto-optical effects in hyperbolic metamaterials based on ordered arrays of bisegmented gold/nickel nanorods. *Nanotechnology* **2021**, *32*, 305710. [CrossRef]
20. Kompaniiets, M.; Dobrovolskiy, O.V.; Neetzel, C.; Ensinger, W.; Huth, M. Superconducting Proximity Effect in Crystalline Co and Cu Nanowires. *J. Supercond. Nov. Magn.* **2015**, *28*, 431–436. [CrossRef]
21. Riminucci, A.; Schwarzacher, W. Coexistence of superconductivity and superparamagnetism in Pb–Co electrodeposited nanowires. *Appl. Phys. A* **2017**, *123*, 161. [CrossRef]
22. Prakash, O.; Kumar, A.; Thamizhavel, A.; Ramakrishnan, S. Evidence for bulk superconductivity in pure bismuth single crystals at ambient pressure. *Science* **2016**, *355*, 52–55. [CrossRef] [PubMed]
23. Ge, J.Y.; Gutierrez, J.; Lyashchenko, A.; Filipov, V.; Li, J.; Moshchalkov, V.V. Direct visualization of vortex pattern transition in ZrB12 with Ginzburg-Landau parameter close to the dual point. *Phys. Rev. B* **2014**, *90*, 184511. [CrossRef]
24. Córdoba-Camacho, W.Y.; Silva, R.M.; Vagov, A.; Shanenko, A.A.; Albino Aguiar, J.A. Quasi-one-dimensional vortex matter in superconducting nanowires. *Phys. Rev. B* **2018**, *98*, 174511. [CrossRef]
25. Córdoba-Camacho, W.Y.; Silva, R.M.; Shanenko, A.A.; Vagov, A.; Vasenko, A.S.; Lvov, B.G.; Albino Aguiar, J. Spontaneous pattern formation in superconducting films. *J. Phys. Condens. Matter* **2020**, *32*, 075403. [CrossRef]
26. Moura, K.O.; Pirota, K.R.; Béron, F.; Jesus, C.B.R.; Rosa, P.F.S.; Tobia, D.; Pagliuso, P.G.; Lima, O.F. Superconducting Properties in Arrays of Nanostructured β-Gallium. *Sci. Rep.* **2017**, *7*, 15306. [CrossRef]
27. Engbarth, M.A.; Bending, S.J.; Milošević, M.V. Geometry-driven vortex states in type-I superconducting Pb nanowires. *Phys. Rev. B* **2011**, *83*, 224504. [CrossRef]
28. Riminucci, A.; Schwarzacher, W. Magnetic signature of granular superconductivity in electrodeposited Pb nanowires. *J. Appl. Phys.* **2014**, *115*, 223910. [CrossRef]
29. Schoeller, H.; Cho, J. Oxidation and reduction behavior of pure indium. *J. Mater. Res.* **2009**, *24*, 386. [CrossRef]
30. Roberts, B.W. Survey of superconductive materials and critical evaluation of selected properties. *J. Phys. Chem. Ref. Data* **1976**, *5*, 581. [CrossRef]
31. Toxen, A.M. Size Effects in Thin Superconducting Indium Films. *Phys. Rev.* **1961**, *123*, 442. [CrossRef]
32. Mason, P.V.; Gould, R.W. Slow-Wave Structures Utilizing Superconducting Thin-Film Transmission Lines. *J. App. Phys.* **1969**, *40*, 2039. [CrossRef]
33. Thompson, R.S.; Baratoff, A. Magnetic Properties of Superconducting Thin Films in the Nonlocal Regime. *Phys. Rev. Lett.* **1965**, *15*, 971. [CrossRef]
34. Lock, J.M.; Bragg, W.L. Penetration of magnetic fields into superconductors III. Measurements on thin films of tin, lead and indium. *Proc. R. Soc. Lond. A* **1951**, *208*, 391. [CrossRef]
35. Dheer, P.N. The surface impedance of normal and superconducting indium at 3000 Mc/s. *Proc. R. Soc. Lond. A* **1961**, *260*, 333. [CrossRef]
36. Guyon, E.; Meunier, F.; Thompson, R.S. Thickness Dependence of k_2 and Related Problems for Superconducting Alloy Films in Strong Fields. *Phys. Rev.* **1967**, *156*, 452. [CrossRef]
37. Fossheim, K. Electromagnetic Shear-Wave Interaction in a Superconductor. *Phys. Rev. Lett.* **1967**, *19*, 81. [CrossRef]

38. Xu, S.H.; Fei, G.T.; Zhang, Y.; Li, X.F.; Jin, Z.; Zhang, L.D. Size-dependent melting behavior of indium nanowires. *Phys. Lett. Sect. A Gen. At. Solid State Phys.* **2011**, *375*, 1746–1750. [CrossRef]
39. Shilyaeva, Y.; Gavrilov, S.; Matyna, L. Melting of indium, tin, and zinc nanowires embedded in the pores of anodic aluminum oxide. *J. Therm. Anal. Calorim.* **2014**, *118*, 937–942. [CrossRef]
40. Kumar, P.; Kiran, M.S.R.N. Nanomechanical characterization of indium nano/microwires. *Nanoscale Res. Lett.* **2010**, *5*, 1085–1092. [CrossRef]
41. Kumar, P. Trench-template fabrication of indium and silicon nanowires prepared by thermal evaporation process. *J. Nanopart. Res.* **2010**, *12*, 2473–2480. [CrossRef]
42. Mondal, A. Synthesis of indium nanowires by oblique angle deposition. *J. Nanophotonics* **2011**, *5*, 053522. [CrossRef]
43. Oh, S.S.; Kim, D.H.; Moon, M.W.; Vaziri, A.; Kim, M.; Yoon, E.; Oh, K.H.; Hutchinson, J.W. Indium nanowires synthesized at an ultrafast rate. *Adv. Mater.* **2008**, *20*, 1093–1098. [CrossRef]
44. Redkin, A.N.; Korepanova, O.S. Spontaneous template free growth of indium nano- and microwires via disproportionation of the InBr aqueous solution. *Chem. Phys. Lett.* **2019**, *734*, 136736. [CrossRef]
45. Talukdar, I.; Ozturk, B.; Flandersa, B.N.; Mishima, T.D. Directed growth of single-crystal indium wires. *Appl. Phys. Lett.* **2006**, *88*, 221907. [CrossRef]
46. Gordon, E.B.; Karabulin, A.V.; Matyushenko, V.I.; Sizov, V.D.; Khodos, I.I. Structure of metallic nanowires and nanoclusters formed in superfluid helium. *J. Exp. Theor. Phys.* **2011**, *112*, 1061–1070. [CrossRef]
47. Li, H.; Liang, C.; Liu, M.; Zhong, K.; Tong, Y.; Liu, P.; Hope, G.A. Synthesis of indium nanowires by galvanic displacement and their optical properties. *Nanoscale Res. Lett.* **2009**, *4*, 47–53. [CrossRef]
48. Chen, F.; Kitai, A.H. Application of indium nanowires to donor-acceptor pair luminescence. *J. Lumin.* **2008**, *128*, 1856–1862. [CrossRef]
49. Liu, Y.; Zhang, K.; Li, M.; Zhao, C.; Wang, X.; Yuan, Z. Ion emission properties of indium nanowires grown on anodic aluminum oxide template. *Vacuum* **2016**, *131*, 209–212. [CrossRef]
50. Watson, J.H.P. Critical Magnetic Field and Transition Temperature of Synthetic High-Field Superconductors. *Phys. Rev.* **1966**, *148*, 223. [CrossRef]
51. Hindley, N.K.; Watson, J.H.P. Superconducting Metals in Porous Glass as Granular Superconductors. *Phys. Rev.* **1969**, *183*, 525. [CrossRef]
52. Roslyakov, I.V.; Gordeeva, E.O.; Napolskii, K.S. Role of Electrode Reaction Kinetics in Self-Ordering of Porous Anodic Alumina. *Electrochim. Acta* **2017**, *241*, 362–369. [CrossRef]
53. Lillo, M.; Losic, D. Pore opening detection for controlled dissolution of barrier oxide layer and fabrication of nanoporous alumina with through-hole morphology. *J. Membr. Sci.* **2009**, *327*, 11–17. [CrossRef]
54. Noyan, A.A.; Napolskii, K.S. Birefringence in anodic aluminum oxide: An optical method for measuring porosity. *Mater. Adv.* **2022**, *3*, 3642–3648. [CrossRef]
55. Tinkham, M. *Introduction to Superconductivity*; McGraw-Hill: New York, NY, USA, 1996; p. 20.

MDPI AG
Grosspeteranlage 5
4052 Basel
Switzerland
Tel.: +41 61 683 77 34

Nanomaterials Editorial Office
E-mail: nanomaterials@mdpi.com
www.mdpi.com/journal/nanomaterials

www.ingramcontent.com/pod-product-compliance
Lightning Source LLC
Chambersburg PA
CBHW062052220926
51850CB00010B/2740
9783725816866